Let's
Talk
Quality

Other McGraw-Hill Titles by Philip B. Crosby

The Art of Getting Your Own Sweet Way (1972, 1981)

Quality Is Free: The Art of Making Quality Certain (1979)

Quality Without Tears: The Art of Hassle-Free Management (1984)

Running Things: The Art of Making Things Happen (1986)

The Eternally Successful Organization: The Art of Corporate Wellness (1988)

Let's Talk Quality

96 Questions
You Always Wanted to Ask Phil Crosby

by

Philip B. Crosby

McGraw-Hill Publishing Company

New York   St. Louis   San Francisco   Auckland
Bogota   Hamburg   London   Madrid   Mexico
Milan   Montreal   New Delhi   Panama
Paris   Sao Paulo   Singapore
Sydney   Tokyo   Toronto

Library of Congress Cataloging-in-Publication Data

Crosby, Philip B.
  Let's talk quality.

  Includes index.
  1. Quality assurance.   I.Title.
TS156.6.C755   1989   658.5′62   88-13524

  234567890   DOC/DOC   895432109

ISBN 0-07-014565-2

*The editors for this book were Jim Bessent and Barbara B.
Toniolo, the designer was Naomi Auerbach, and the
production supervisor was Suzanne W. Babeuf. It was
composed by the McGraw-Hill Publishing Company
Professional & Reference Division Composition Unit.*

Printed and bound by R. R. Donnelley & Sons Company.

A sixty-minute audio program to accompany this book
is now available. Ask for it at your local bookstore
or phone toll-free 1-800-2-MCGRAW.

*For more information about other McGraw-Hill materials,
call 1-800-2-MCGRAW in the United States. In other
countries, call your nearest McGraw-Hill Office.*

Contents

Preface   vii

Quality: The Way We Were   *1*

The Quality Revolution at Home and Abroad   *11*

Quality Thinking: Toward a Mature Philosophy of Quality   *43*

Quality Action: Toward a Workable Quality Process   *99*

Quality Relationships: Toward a Quality Business Culture   *137*

Epilogue: Quality Past, Present, and Future   *175*

Guidelines for Browsers   *183*

Index   203

Preface

When I started Philip Crosby Associates in 1979, I soon realized that the educational part of the quality improvement process aimed at executives and managers had to be intensive. People just weren't going to learn in a hotel ballroom with several hundred of them involved. We were dealing with "brain damage" resulting from years of ineffective information. The whole approach had to be more personal.

I felt that if classes were limited in size to around 20 and held continuously for a week, this would allow a relationship to develop that would ensure the students' comprehending the concepts required. As it developed, we ended up with 22 per class and still do it that way in all our operations, domestic and overseas. It has worked out well.

In the early days I taught all of the classes myself, but as the business began to grow, this became increasingly impractical. After developing course materials and a few videos, with the help of Bob Vincent I began to teach others how to conduct the courses. It takes about nine months to learn to handle a Management College course and longer for the Executive College. All courses have full-length filmed case histories now.

As my personal participation in the classes diminished, I began dropping in on each class for a period of "open discussion." The idea was to give participants a chance to ask whatever was on their minds and to smooth the way for the instructor by getting some things settled that might arise. Besides, I really enjoyed it. It was also very helpful to me in developing new material. This way, I could keep up with what the managers and executives on the front lines of business were thinking.

As the class load became higher, the college president diplomatically hinted that the question and answer periods were becoming a bottleneck and suggested another way. Now, the Winter Park classes all have lunch together and then gather in the auditorium for one big question and answer session. We also began to tape the sessions so that people at other PCA locations could share in what was being said in these open exchanges. Subsequently, it was decided to offer the tapes as a subscription set. The tapes are a way to bring anyone into the circle and to clarify for them the quality philosophy and methods. They let people know what I am really trying to say rather than what people think I am saying.

When I learned that surgery was going to put me on a short leash for a few weeks, I began to think about making a book out of some of the taped material from the question and answer sessions. It was good therapy. I learned a lot in the process, and the book is now at hand.

Whereas at first I thought such a book would be a simple matter of editing my remarks as they were on the tapes, I learned that the exact written transcript of a verbal presentation does not read well at all. Still I tried to keep with the original ideas expressed in the questions and to use many of the ideas contained in my original answers, although I have broadened my responses and in many cases have refined them.

The questions change over the years, but the majority

have been consistently oriented toward the concepts that underlie our philosophy of quality and toward specific problems experienced by participants of the Quality College in getting things going back home in their respective businesses.

The questions and answers selected for *Let's Talk Quality* fall quite naturally into five areas.

The first part, The Way We Were, serves as an introduction to all the rest and presents a picture of how far things had degenerated. These four questions and answers deal, in part, with the conditions and thinking that led to what might be called a "quality crisis." They represent a good starting point—and a basis for comparison—to show the directions we have tried to go, to show some of the false starts, and to set the stage for presenting the future agenda.

The Quality Revolution section contains questions and answers that deal with changes on the domestic and international scenes. It should not be surprising that much of this discussion deals with the Japanese, their competitive position, and their approach to quality as compared to that of the United States. Many of the questions relating to the domestic scene are concerned with the nation's largest, and therefore most visible, corporations. How the giants cope with quality issues is understandably a model for how other businesses will cope.

Part 3 on Quality Thinking consists of questions and answers that shed light on the quality improvement process itself and the thinking that underlies it. Ours is a changing, growing philosophy, but it is built on principles that remain stable. There is material here on the 4 Absolutes, the 14 Steps to quality implementation, the concept of zero defects, and the thinking behind the belief in total conformance to requirements. In addition, Part 3 deals with the relationship between quality concepts and implementation and with other quality and productivity approaches such as MBO and Just-In-Time. There are questions dealing with subjects as all-encompassing as how organizations change and as detailed as inquiries about the hows and whys of filming *The Quality Man.*

The section on Quality Action makes clear that problems cannot be solved in an ivory tower. Philosophies must be executable, just as actions must be guided by philosophical principles. These questions and answers involve the problems that attend implementation. All businesses are different, and in this section you will hear from representatives of high tech, retail, and service businesses, of large businesses and small businesses, and from people in entrepreneurial situations, all posing tough questions about the nuts and bolts of quality implementation.

Finally, the fifth section, Quality Relationships, revolves around questions about quality as it relates to business culture. Business can be seen as a set of "people" relationships. Any attempt to change business must take into account the relationships between and among management, employees, suppliers, and customers, for they are all equal partners in the enterprise. Despite the best intentions, there is no denying that change generates anxiety, and the "company politics" that result can be woefully counterproductive.

I've concluded *Let's Talk Quality* with an epilogue that attempts to sum up where we've been, where we are, and where we want and need to be in the quality quest. The 96 questions and answers in *Let's Talk Quality* represent the beginning of the kind of "quality dialogue" that I see as essential to an ongoing quality improvement process. They are by no means all the concerns voiced, but only a selected sampling. You may think of other questions. If so, I invite you to commit them to paper and send them to me. Now, let's talk quality.

Philip B. Crosby
Winter Park, Florida

Quality: The Way We Were

Question 1

You've been quoted as saying that in the past five years you have learned more about the quality process than you had in the previous 25 years. What five areas on the quality front have seen the most dramatic change in the past five years?

Answer

I think everyone learns more in the past five years than in all the previous years in any field. But my exposure during that time has certainly been extensive. In quality I think I have learned more about several things.

First is implementing. We have learned how to teach companies how to do for themselves rather than doing it personally for them. Years ago, I used to go into an operation, figure out what the problem was, and then guide the company along. If you read Tennant company's book, *Quest for Quality*, you will see that the help I gave them was on a personal basis. We had no company then, no products, no tools, only our first management classes in the Quality College. I taught them the basics, gave them the benefit of my experience, and they carved their way through the jungle. If they had had available all the tools we have today, they could have flown over the jungle instead of walking through it.

Tennant had to educate all their people; we can now supply that. They had to go out and find or create tools; now these same tools are wrapped up in software and other media. It could have saved them a lot of time, money, and effort.

Second is attitude. Attitudes change when a business's culture or working environment is changed, not until. Getting people together and preaching to them, or "motivating" them, changes very little. A person has to experience the value of a new look at life. Even couples who live side by side can have vastly different attitudes about their apartment building, their marriage, and everything else, based entirely on their experience. When it is a pleasure to come to work because the requirements for quality are taken seriously and management is helpful, then attitudes change permanently.

Third is the CEO. When it comes to changing a company, I have learned that someone has to sit down and talk to the chief about the extent to which he or she understands the personal role in making quality happen. No one inside the company is going to do that successfully. It has to be an outside party. That's why someone senior in an organization like ours has to have a planned session with the CEO. Then we can tell them the truth.

Fourth is wellness. In my most recent book, *The Eternally Successful Organization*, I talk about causing "corporate wellness." The message is that there is no necessity for companies to have all the problems they have. They don't need to be subject to illness, but they do nothing to prevent it. So instead of just talking about prevention in terms of product or service or administrative problems, it is best to talk and think about it from a corporate standpoint. A corporation is like a person, and what happens in personal wellness is a great analogy for the business world. And this involves taking a holistic view of the body, whether it is the human body or the corporate body.

Fifth is witnessing. I learned, even though I had been preaching this for years, that when it comes to quality, the witness of management is more important than anything else. Teaching people, leading people, showing people, providing tools—everything loses meaning if employers, customers, and suppliers feel that management is not walking like they talk. Look around. If you'll notice, our associates here at the Quality College, for example, believe that their management is serious about a quality environment. Therefore, everyone pitches in to keep the place immaculate, and they are vitally concerned that visitors and students have an enjoyable time. Everything runs on time because we all want it to be that way.

That was a good question. It made me think of things in a way I had not done before.

Question 2 _____

Are there not, in fact, some situations in which it is cheaper and more efficient in the long run to manufacture a product, monitor the results, and sample to get the ones that you want, rather than to slavishly attempt always to do things exactly right the first time?

Answer

The philosopher George Santayana once said, "Those who do not remember the past are condemned to repeat it." What you just described is the system that got us into a quality fix in the first place. People were making decisions like that all day long, and as a result the product completely lost its integrity. We never did learn how to develop processes with high yields and systems that were reliable.

Sampling, for instance, is based on having a homogeneous population. Then a randomly selected few represent the whole of the box, or barrel, or tub, or whatever. However, when repeatable processes are never established, then sampling itself becomes very chancy and unreliable. Now if a product is sampled and the sample is found to contain too many defectives, what happens? A bigger sample is taken, and so on and so on until the lot is sorted. However, if the sample passes, the lot is approved for the next step.

When management continually interferes with the work process by making value judgments based on assumptions of whether it is cheaper to do something this way or that, then the process never becomes mature. I have had a standard bet for years that I would pay $100 cash to anyone who could prove to me that it was less expensive to do something wrong the first time. No one has ever collected. The notions you described are part of the mythology people pick up along the way as they learn management.

People should spend their time improving the quality process, rather than juggling it around to meet their feelings of

the day. Each step along the way needs to be continually examined to see if it can be done cheaper, quicker, more reli-
ably. Don't set up false detours or special arrangements.
Learn what right is and do it that way all the time. Then people will have something they can trust.

I had a car once that seemed to start only when it felt like
it. It had an automatic choke which would not do its job on a
regular basis. It was necessary for someone to open the hood
and hold a little lever down while another person started the
engine from the driver's seat. When the young lady I was dat-
ing told me she was not going to do that anymore, I reluc-
tantly went to the garage to get it fixed. I knew it was going to
cost a lot less than getting a more cooperative girlfriend. However, as it turned out, only a minor adjustment was involved
and the mechanic enjoyed my predicament so much he didn't
charge me.

When management encourages procedural Band-Aids, employees lose confidence in them and in the process. Don't
outsmart yourself. Legitimate trade-offs are few and far between.

Question 3 _____

*We often hear the phrase "planned obsolescence." How does
your vision of quality improvement approach that concept?*

Answer

This concept was developed by Raymond Lowrie back in the
1940s. It is misunderstood, like most concepts are when they
begin. People think it means that the product is designed to
fall apart the day after the warranty expires, or that it should
only work to a minimum standard so the customer will get
tired of it and buy into the succeeding generation.

What planned obsolescence really refers to is the practice of
continually offering new and more attractive designs or per-

formance characteristics so consumers will yearn for the new model. They will want to go buy the latest attraction. From this thought, for example, came fins on automobiles. Indeed, most of the design alterations that we continually see on products, and services too for that matter, are the result of this idea.

The quality improvement process, as I conceive it, handles all of this well. We deal with change continually. Therefore, change should be a friend. It should happen by plan, not by accident.

For example, a couple of years ago, our company began an entirely new way of dealing with quality assurance, change control, internal auditing, and other appraisal systems. We call it *systems integrity*. This concept is introduced and explained in detail in *The Eternally Successful Organization*. Briefly, the idea is to make a friend of change by doing it on purpose in a way everyone knows about. At the same time, the systems integrity people are continuously examining the systems of the company to see if they are operating properly. They do all this appraisal without giving advice. That cuts out the business of having one group of experts that says what is okay and what is not okay.

Question 4 _____

The concept of zero defects has actually been around for a long time. Why do you think it didn't catch on initially? Why it is just now coming into vogue?

Answer

There is a whole saga involved in the answer to that question. I will explain it as best I can with the background that good ideas based on solid concepts have a great deal of difficulty in being understood by those who make a living doing things the other way. Dr. Joseph Lister, for instance, who was the first one to try to get doctors to understand that diseases could be transmitted by unsterile conduct, was driven insane.

The idea that doctors themselves, who would blithely wipe their bloody hands on their coats, were carriers of illness was too much for the profession to handle. It took years before anyone would even listen. Granted, Lister didn't have a very good approach. He attacked the people rather than the problem. Instead of yanking doctors away from their patients, he needed to offer proof of the problem and of his proposed solution. It took another generation for physicians to understand asepsis.

I am hardly in the league with Lister, but the events in the quality field run parallel in many ways. My idea back in 1961 was that people were receiving performance standards in business that were less stringent than the ones in their personal lives. They expected to do things right when it came to holding babies, paying bills, and coming home to the correct house. In business they were given "acceptable quality levels," waivers, and deviations.

When I announced the concept of zero defects—and it is all spelled out in *Quality Without Tears*—everyone leaped on it as a motivation program for workers. The Department of Defense and my employer at the time, the Martin Company, held seminars to promote it and spread it around. It went through the defense space industry in a hurry. Within two years everyone was conducting a "Zero-Defects Program," with a capital *P*.

I kept trying to explain that this was a concept that *management* had to do something about, but to little avail. No one would really listen to me. The quality professionals denounced it, and the consultants and teachers of the profession included explanations of how foolish this was in all their speeches and articles. When I suggested that their hands were bloody because they had institutionalized defect rates, I was judged to be mad. I was angry, perhaps, but I was not mad. None of them bothered to probe for my real meaning.

While all this was going on, Dr. Kobyashi of Nippon Electronic Corporation (NEC) came to see me in Orlando. I explained to him what I really meant by zero defects and when he returned to Japan, he set it up. Last summer the Japanese had a celebration of 16,000 people representing 3750 companies, recognizing 20 years of zero defects in Japan.

In the United States the zero-defects idea hung on over the intervening years and many companies applied it properly. I get letters regularly from organizations that have operated by the zero-defects principle since 1962 and have had great successes. Good ideas don't die.

Since 1979, when *Quality Is Free* came out, zero defects has been receiving the attention that comes with understanding. Executives, who know only that what they have been doing all along is not effective, understand that zero defects is another way of saying "do it right the first time." For them, it is a way of communicating what they wanted all along but couldn't get people to understand.

The quality professionals and their teachers still have a problem with zero defects because some things are deeply ingrained and deeply confused, for instance, the Pareto principle that says 80 percent of all problems come from 20 percent of the causes. This is true in almost every situation. Likewise, 20 percent of the people pay 80 percent of the taxes.

However, what happens in traditional quality control is that a division is made of nonconformances into the "vital few" and the "trivial many." The trivial many are not dealt with; they are left to poison the product for the customer. It is felt that there is not time to work on them. In a true zero-defects approach, there are no unimportant items.

These ideas have to be put into their proper perspective. Analytical tools have to be used to find out where we are, not to justify error.

Between 1962 and 1979, when I went into business on my own, I made more than 1500 speeches, wrote dozens of articles, gave a couple hundred interviews, and published four books. I never missed a chance to explain the concept of zero defects properly. To all but those who suffered permanently frozen opinion and didn't listen, it began to make some sense. However, it was difficult. Moderators would thank me for my speech and then assure the audience that I "didn't literally mean things could be defect-free."

Once one of the "grand old men" of the quality profession came up on stage out of the audience to ask me how I could consistently defend such a stupid thought. "You seem bright enough," he said. "You probably have never listened to what you are saying."

So I asked him to explain to me what he thought the concept of zero defects meant. He talked about worker motivation, how the idea of exhorting the worker to do better was useless, and how it was causing people to not use the real tools of quality control. I walked across the stage, put my arm around his shoulder, and said, "If that is what zero defects is, I don't want any part of it either." That had no effect on him. To this day he still doesn't understand that we are on the same side.

The zero defects idea never died. It just had a long gestation period. Whereas people used to think I was exaggerating for effect, now they feel that I am saying it like it is. Any other performance standard is not worthy of our effort.

I have learned that whenever I come up with what seems like a good idea, it is an even better idea not to spread it around until I know how best to explain it to people. You can't ignore history. No idea was ever accepted right out of the chute. There is no reason to think that mine should break this mold.

*The
Quality
Revolution
at Home
and Abroad*

Question 5

What is the status of the so-called quality revolution in this country? Are we still gaining momentum, or is the effort tapering off? What do you predict things will be like two years from now?

Answer

Things are improving and that improvement rate should increase rapidly over the next few years. It could be doing much better but it is hard to change business philosophy. There has been a lot of effort expended on quality over the years but most of it has not achieved much. People work on the wrong things; they aim their activity at the bottom of the organization.

Back in the 1970s, as ITT began to be known for its quality efforts, there was a continual stream of visitors to its offices in New York. Two or three people from different companies would come to learn what was being done, and we would explain it to them.

Tim Dunleavy, the ITT President, asked me one day if we should really be doing this. "Won't these people take what you have developed and then compete with us better?" he asked.

"No way" was my comment. "The key to quality is having management understand that *they* are the problem. How are these people going to go back and say that? Are they going to write a report saying that their president never saw a waiver he didn't like, or that the chairman has the ethics of a burglar? They're not going to do that. But if they start sending the chairman over to find out what we do, then I will put a lid on it." He agreed.

Inside ITT the management readily took the responsibility for quality. No one spoke of "quality problems," for instance. They would talk about "marketing problems," or "design problems," or "service problems." That way the right people went to work on it. When the Japanese started showing American products up a few years ago, all the quality experts went to Japan and came back with "quality circles" and "statistics." Both of these are valuable techniques and should be used. They are widely available. For example, a course called statistical process control (SPC) is taught at the Quality College and there is software for it. And worker communication groups resembling quality circles have been part of my writing for a long time.

However, quality circles and statistics are not the be all and end all of the world. They are only a small part of the task of *causing* quality. To use only two tools, neither one of which requires management policy changes, is naive. It is like playing golf with a five iron and a wedge. Resorting to panaceas has held the quality recovery back because management thinks something is happening and very little is. Meanwhile, the rank and file figured out very quickly that management was not truly changing its wicked ways so they became discouraged and quit trying. Then everyone goes looking for another tool.

The past few years have seen much more interest in quality on the part of senior management. The typical group that comes to us has tried everything they read about and decided that nothing is changing. We work with them to alter their culture and install the basic concepts that make quality a friend instead of a liability.

Now more and more senior people are making speeches about quality, and there is a nationwide "quality month" and a lot more activity. Unfortunately, most of it is still aimed at "goodness" and is rather vague. But it is a beginning. For some reason, our brothers and sisters in the quality profession, at least the ones who have been at it for a long time, have been very reluctant to help management do what it needs to do. I guess all professions are loath to relinquish any of their mystique.

In time, the results obtained by those who are successful in causing quality improvement will bring others along. But the nation as a whole has an enormous way to go. I think about one-third of the companies are working on the right track, one-third are working just as hard on the wrong things, and one-third are hoping for a law or something to come along and save them.

Question 6 _____

In your experience, how do organizations change? Do they go through recognizable phases, or does it all seem to happen at once?

Answer

There seem to be three phases in change, and they apply in personal life as well as in business. Quitting smoking is a good parallel example, because smoking is not a medical addiction. It is a habit, a cultural thing. (I should know, I did it for 20 years.)

The first phase of change is developing conviction. That is when a person or organization's leadership decides that the problem is real and that it is time to do something about it. This conviction could be because we are tired of hearing the family or the doctor complain about the "bad habit" and its effect on them. In business it could be that customers are leaving because the product they are being given does not perform properly. Whatever the reason, the first phase is characterized by a determination to do something. One of the first actions is to start to study the situation and poke around to see what other people are doing about all of this. For example, on any given day at PCA there will be at least one management group visiting to see what we can do for them and how we would do it.

The second phase is commitment. Most people think this is the end of the road, but it is really the beginning. Here is where the ex-drinker, for example, is saying that there will be no drinks that day. The ex-smoker is passing up cigarettes for an hour. And the group that waivers everything is tearing up forms, cold turkey. Commitment requires the demonstration of seriousness by doing, or, in some cases, not doing. Everyone is watching, hoping, and wondering. They are anxious to see what will happen when a real test of the commitment comes. Will the newly committed weaken? Will stress cause cracks in the quality policy? Will anyone be around for the day when zero defects becomes reality?

The third phase is conversion. By this time, it would take a gun to the temple to cause the ex-smoker to smoke. It would require the threat of assassination to produce agreement to use anything that is not in conformance with requirements, but no one will even ask. The converted are not tempted by the desire to take shortcuts or to go back to destructive ways. They begin to wonder what all the problem was about in the first place and why it took so long to decide to change to a new way of life. Life seems so much easier now. The converted stay converted.

Question 7 _____

Do you think there is a reason why there aren't more food businesses involved in the quality improvement process?

Answer

There actually are quite a few, like Bama Pie, which produces all the pies for McDonald's, Perdue Chickens, and several others whose names can't be mentioned. But it has only been recently that the food industry has bestirred itself in terms of quality improvement.

There are several reasons for this. Food people think they are very quality conscious and do not see much need for change. Actually, since there has been no significant foreign competition, they have had only each other to deal with. Now much of that has altered.

Another reason is that there are some really weird specifications in food technology. You can do what the specifications say and be considered a wonderful supplier. This has always bothered me.

Back in 1977, when I was still with ITT, I spoke to a large session of food quality people. My presentation was at the luncheon session. Prior to that meeting, I toured an exhibit and came upon the specification for foreign material in flour and other commodity supplies. So at the luncheon I announced that I had examined the flour used to make the rolls and could assert that there was no rat feces in it. However, since the spec permitted a certain amount, I noted that we had taken action to add some in order to conform to the specification. They were not amused.

The point here is that laying out a specification that says something undesirable is acceptable does not make it right. The requirement has to be something that delivers to customers what those customers think they are buying. Hardly anyone at that luncheon appreciated my comments, but now I think they are getting the message. They are taking quality more seriously.

The purpose of quality is not to accommodate the wrong things. It is to eliminate them, to prevent such situations. It is the very same thing as the soldering defects I encountered when working at Martin [see Question 22]. As long as we had an acceptable level of them, they existed. When they became unacceptable, they disappeared.

I see the food industry changing. The same needs to happen with airlines. They think you are the problem, not them.

Question 8 _____

Is it true that the airline industry has shown little interest in your brand of quality improvement?

Answer

There has been some interest from airlines recently, but it is at a low level and is very cost-oriented. The consultation necessary for effective quality improvement may sound expensive, but you must consider what you're getting. Also, the more people a company involves, the lower the cost, relatively. I wouldn't want to deal with a company that isn't going to involve all of its employees.

The problem with airlines is that the management doesn't think any of the quality problems are their fault. They blame the passengers, the airport authority, the FAA, the unions, the aircraft manufacturers, each other, and even me for raising people's consciousness about quality. It is hard to deal with that attitude. I tell them they need a better class of customer, one that doesn't complain.

Running an airline involves a lot of people. However, none of them are against air travel. They all want it to come off well. Everyone is for on-time operation, safety, fair wages, proper fares, return on equity, and a happy life. Yet in trying to accomplish this, they somehow encounter constant wrangling and hassle. It is hard to complete a trip without at least one incident that brings even the calmest close to rebellion.

The reason for this is that there is no clear set of requirements for operating the air-travel system. No one can be in charge of the entire thing, but everyone should be able to agree on who is going to do what. Management thinks that the customers are only interested in ticket price, which is not true. Customers think the airlines are only interested in making money, which is not entirely true either. And the government regulatory agencies think everyone hates them, which probably is true.

In the prevention implementation courses offered at PCA, we use a fictional airline, Global, as a case history. A two-hour film was created around an airline's problems with baggage handling, and the course of events runs parallel with the hero's wellness battle. While the airline is getting its problems solved, the hero loses weight, quits smoking, and gets back in shape. The solutions we invented for this simulation problem turn out to be what they have been thinking about in real life.

Question 9 _____

Isn't it true that the kinds of problems some of the major man-
ufacturers, such as General Motors, are having now are not so
much in the production area as elsewhere in the organization?

Answer

The problems every manufacturing company has are mostly
not in the manufacturing area. They originate with the marketing, engineering, management—in all the places that
think it is okay to make things up as they go along. Now the
manufacturing people make a lot of problems on their own,
like not training employees how to do the job, or having bad
supervision, or not listening to the rank and file. There are a
lot of problems that can—and usually do—happen.

But when you have an environment where nothing is certain, where almost anyone can make changes or write deviations, then the output is uncertain at best. I had a senior executive at one company tell me that their biggest problem was
basic design. I suggested that they had never actually made
anything like the design and, therefore, they had no idea of
whether the design was any good or not. Think about that.

General Motors, like most large manufacturing companies,
is very difficult to change. There are entrenched ideas and
concepts all over the place. The fact that the traditional quality control techniques have been part of the problem over the
years does not let them become any more interested in changing things. It is hard to get people interested in improvement
of any kind if they perceive it as a threat to their authority or
lifestyle.

Companies like this are like the Mississippi River. They
have grown accustomed to tossing the idea for a new product
into the stream up at the source, and as it floats down, each
functional area does its bit. Finally, when the product arrives
at New Orleans, it is complete and ready to sell. The bad
news is that no one is really responsible for it, and no one
owns up to causing problems. So the entire organization

learns how to adjust. It becomes very adept, very resourceful at making things "come out right." Soon, half the operation does nothing but pick up after the other half.

A good mathematician can figure where the money goes in such a situation. But unless management is really ready to re-educate everyone and restructure the responsibilities so they are clear, their future is dim. When I was in the Navy, I had to stand by my bunk with my gear all spread out in a pre-scribed manner, and the chief would come by for inspection. If anything was not like it was supposed to be, there was no one to blame but me. I knew that, he knew that, it was agreed. It isn't like that in big companies, and they suffer for it. However, none of them are as big as the Navy, so they could change if they wanted to.

But General Motors is improving rapidly, even if it does tend to do many things the hard way. It has redone all of its plants. It has become adept at project management and is be-ginning to pay better attention to workers and customers. The GM-10, Truck and Bus, and Corsica/Beretta projects have all done very well. The component divisions have come alive and are supplying defect-free material to their sister divisions. GM has spent 40 billion dollars getting well. It probably could have bought the entire Japanese auto industry for four billion back in 1971.

Question 10 _____

What is your favorite example of a company that turned things around as a result of adopting the quality improvement process?

Answer

There are a lot of good examples of people overcoming quality difficulties. Sixteen or so of these have been documented on vid-eotapes as part of the "Quality in the 21st Century" project. We call these tapes Q-21s. In these tapes the individuals inside a given company talk about their experiences. It is very interesting. The tapes are used to help encourage those inside a company that is setting up a quality improvement process.

Companies that I am especially proud of make up a good-sized list. I have given "Fanatic's" awards [see Question 51] to Perdue Chickens, Southwest Bell, Chevrolet's Corsica and Beretta, Armstrong World Industries, General Motors Truck and Bus, Milliken, and several others. IBM and Tennant are two companies that adopted the process very early and have done very well. Banks, retail stores, and many other businesses have well-documented success stories.

One story that holds particular interest happened inside General Motors. There is a two-tape Q-21 on it. It involves what they then called the L car, which became the Corsica and Beretta, two of Chevrolet's newest vehicles.

When this project was laid out, GM decided to make it a team effort. So they brought together two assembly plants, Wilmington, Delaware and Linden, New Jersey; two pressed-metal plants in Ohio, one at Parma and the other at Mansfield; along with the engine plant at Tonawanda, New York. A project office was set up at GM headquarters to coordinate, and the engineering department was brought into the team.

Now this was a brand new idea for the 75-year-old company. One of the requirements for team membership was that everyone had to attend the management or executive college here at Winter Park. Then all the employees had to go through what is called the quality education system or the quality awareness experience.

One of the executive college participants, the manufacturing director, went to the school and didn't like it. He returned home and wrote me a letter. I answered, he called, we talked, and he came back with his team. This time they realized that the objective of the management and executive college courses is not to provide them with a pill. This was something

they had to do themselves. They decided that this was one automobile that was going to be made with zero defects all the
way through. The team established regular meetings, rotating
from plant to plant. I was invited to attend many of these and
had the opportunity of meeting many employees working at
different jobs. The participation of the bargaining unit officers
was solicited. This "jointness" had a great deal to do with the
success of the project.

One of the general managers said he had been with GM for
34 years and this was the first time he had sat down with other
general managers unless there was a battle to be fought. They
all got to like each other and were determined to properly support the L car. While this was going on, the operating management as well as union executives were going through the
same courses of instruction, and other employees were going
through the quality education system and the quality awareness experience. Most important, the words of quality understanding were becoming standard usage. People didn't fight
about the degrees of quality; they talked instead about requirements for quality and whether they were met or not.

Meanwhile, several GM managers spent six or seven
months to actually learn how to teach the management college themselves. Then we helped GM set up their own insti-
tute. We would provide the integrity management, all the
material, and oversee the operation. But they ran it, and today 88 managers a week go through there.

When I would visit a plant, we would use the occasion for
team building. It also offered me the opportunity for personal
evaluation of how things were coming. The format for a visit
became fairly standard. The general manager, some key management people, and the union committee would meet me
and we would go on a tour right away. I like to move around
an organization and talk with people. That way I can get a
feel for what is happening.

Since most of the people had been through our courses,
many recognized me and waved. Then I would go over and

chat. I always asked, "What's your biggest problem?" This question never fails to bring forth some great answers. Also some of the small groups working on special projects or defect eliminations would make presentations as we arrived at their stations. The office areas did the same thing.

Then we would usually go to a conference room where the team would make a 45-minute or so presentation on their status. After that there would be lunch, a couple of speeches by me to large groups, and then a wrap-up in the conference room. At that time they would ask me to tell them frankly how I thought they were doing. And I would tell them—frankly. It all made for excellent communication and we all grew from the sessions.

For me, the wonder was watching these operations change from visit to visit. A couple of them went from dark, dull places where no one worked with each other to being bright and full of camaraderie. In every case there was a transformation, due primarily to everyone understanding quality the same way.

Now, defect-free parts arrive on schedule, there is virtually no rework after or during assembly, there is no salvage area. And the Corsica and Beretta made up the best "launch" in GM history. The same results were obtained in the Truck and Bus Group. In both cases, the management really got in there and participated.

They used the knowledge they had been taught and the tools supplied to them. But unlike many management people, they did not look to techniques as a substitute for good manners and compassion in their communication.

Question 11 _____

Would you continue the story with respect to the other automotive manufacturers, Ford and Chrysler, which most people tend to acknowledge have done a far better job sooner of im-

proving their reputations for quality? Do you agree with that, and if so, how did they bring it about?

Answer

I never judge a management because I know from my own experience that sometimes even doing the right thing turns out wrong, and it is very easy to second guess what they did and didn't do. Having said that, let me take a quick trip down memory lane concerning the automobile industry. It probably directly affects 25 percent of the jobs in our nation and indirectly affects many more.

Back in the late 1960s when I was at ITT, I was, as noted before, making a lot of speeches about the need to improve quality in America. I remember being severely criticized for saying that if we didn't change, one day soon the worldwide definition of tacky and unreliable would be "made in the USA." One area that would not talk with me at all—and it was free then—was the automobile industry. The only one to show any interest and actually do something was Pontiac, which was then run by John DeLorean. He asked me to talk with his management. I did and they started an improvement program based on zero defects. It wasn't run properly, but it still did a lot of good. It was killed as soon as he left.

At the same time two other things were happening. I won't get deeply involved in discussing them, but for those who are interested in the comparative stories of Nissan and Ford, I would suggest reading David Halbersham's book *The Reckoning*.

So, while I was trudging around with my sign saying "Repent, the end is near" concerning quality, there was a group of oil economists who were trying to get the auto people interested in the future price of gasoline. They tried to get appointments with someone important in all three companies and failed. Their message was that the middle eastern rulers were beginning to realize how much money they were giving away and that the United States was living in fool's paradise.

I never met these oil economists, but we were getting the same reception. The reason for this was that business was wonderful. Big cars were selling like mad. The Japanese were just getting started, and their vehicles were viewed as toys. Early Japanese efforts to export into the United States were clear disasters. The cars would not stand up to the American long-distance driving habits and in general were unimpressive. After all, it was impossible to drive for several hours at 65 miles an hour in Japan. One would fall into the sea rather quickly.

However, in 1973 the oil shock came. Suddenly gasoline was a dollar a gallon and scarce, as opposed to 32 cents a gallon and flowing freely on every corner. Toyota and Nissan, who had lots bulging with unwanted, finely made, high-mileage vehicles suddenly couldn't get enough of them. The Big Three suddenly had lots full of large automobiles and no production capability for small cars.

While the American car manufacturers screamed for government protection, the Japanese gained long-time customers who were delighted to learn that a new car did not automatically come with a dozen hidden defects for the customer to find and get fixed.

As what was happening began to sink in, the manufacturers took several different routes after a few years. General Motors decided that its way of operating was the right way and what they needed was new equipment, higher technology, and a new organization. So they spent 40 billion dollars doing the whole company over. They have attacked quality earnestly but inconsistently and are gradually working their way out of the quagmire. They have lost a great deal of their market share, going from 52 percent to 32 percent depending on how one counts such things. But it has definitely seen a big drop.

As they have traveled this route, they have shrunk the company a bit at a time, being criticized for it along the way. They almost ran out of cash once and had to sell a lot of property, including their headquarters. However, now they have a lot of cash and, while still losing market share, have at least stopped the bleeding. They are getting their act together. Today their quality is as good as anyone's.

Ford recognized the inevitable early and restructured the company dramatically. I don't know the exact numbers, but they cut it down by about a third, taking all the heat at once. The new CEO, Mr. Peterson, announced that engineering could design the cars they and marketing agreed on, thus shutting out the finance department in such matters. They also put an emphasis on quality and getting things done right the first time. They insisted that their suppliers do the same, and management treated it all very seriously. Then they advertised what they were doing to help people feel their results would be worthwhile. Most of their improvement has come in the factory, which may haunt them later.

Today General Motors, Ford, and Chrysler have essentially the same quality levels—very much improved over a few years ago—but the public feels Ford is the leader because of what Ford has told them. The Japanese are just a tiny bit ahead and the Germans are in the rear.

Chrysler went through the most dramatic changeover of the Big Three. They came back from literal, if not legal, bankruptcy and are now building fine cars, primarily in the small and medium-sized range. Recently they moved into luxury cars. Their company is a fraction of the size it was before the oil and quality shocks. However, they are the most world-market oriented. They import cars from Japan and have agreements with other international operations to replace the overseas activities they had to sell in order to survive.

Chrysler has had the advantage of being viewed as a scrappy underdog, while General Motors has had the burden of being considered arrogant and uncaring. In reality, there is not much difference between them, but perception is everything and General Motors seems to be snakebit in that it has a persistent habit of shooting itself in the foot.

My dad used to say about boxing that a good big boxer will beat a good little boxer every time. Certainly they don't all turn out that way, but it is the smart bet.

When it comes to quality, the company that teaches its management about prevention and makes that a vital part of the day-to-day upper-level conversation of the company will forge ahead eventually. Both General Motors and Chrysler have set up schools to do that. Eighty-eight managers go through the GM school each week. Chrysler does the same but with a lower number, of course. The 4 Absolutes [see Question 22] are a normal part of operating in all areas of the companies. Regardless of what techniques they use or do not use, quality has become an integral part of the management style and the company culture. It is in their blood.

The message of all this is clear. Listen to everyone, seek out your customers and interrogate them. Don't fall into ruts. Do what you do thoughtfully, but don't go to sleep. Someone may be out in the garden digging up your treasure.

Question 12 _____

Would you say a little about your experience with the Senate oversight committee on NASA?

Answer

Actually, it was a House committee run by Representative Bill Nelson, who had been a passenger in the flight before the Challenger disaster. That flight was found to have aborted for the wrong reason, which may have saved it from disaster by mistake. The committee was assigned to investigate the Challenger disaster, particularly from the perspective of whether the quality control system worked or not.

I told them that in my opinion the quality control system was basically acceptable; the problem was that there was not enough attention paid to it. In fact, it was reporting organizationally to the engineering department. The chief engineer of NASA was brought to the session to defend it. One of the senior NASA people had said a week before something about quality being like caviar: A little on some toast is wonderful, but too much will make you sick. I thought that was a dreadful outlook. It made me feel the whole outfit was sick.

What I said to the committee was that a company could have a wonderful accounting system, but if management didn't pay any attention to it, they could run the company into financial oblivion. This is what had happened to quality control, which, even though acceptable, was not all that wonderful in NASA's case.

Those of you who have been in the aerospace business know that you never put development people in charge of production. They get too used to changing things whenever they feel it is necessary. That is what development is all about. If this pulse isn't quite right, let's try another one until we get it right.

When a twice-a-month schedule comes along, the easiest way to keep things moving is to write waivers, accept deviations, make changes. And that showed up very clearly when the executives decided that the engineers were wrong and certain components could indeed function at a colder temperature than thought possible. As it turned out, the engineers were right.

What I said was that the senior executives did not have the right to overturn things when quality control said it was not according to the requirements. I feel that they plowed through in order to meet schedule and that the tragedy was unnecessary.

But the basic problem in my mind was that NASA had let quality decline over the years. They had rearranged it organizationally, they had denied it access, they had given waiver authority to people who had no business having it. Back in the Apollo days, when quality said no, that was the end of it. In the earlier NASA programs there was a Quality and Reliability Lab in Huntsville, Alabama. They called the shots. When the director, Dr. Diter Grau, left, it was pulled apart.

But quality is not a matter of having some superknowledge in some supergroup. It is a matter of managerial integrity. Either the requirements are taken seriously or they are not taken seriously. In aerospace they are not taken seriously enough, and in NASA I feel they have been treated as guidelines. That is one of the reasons everything is so expensive and very few things work when they are received.

Incidentally, all of what the committee heard has made little difference. I even volunteered to serve as an adviser to NASA, for free, but have never heard from them about it.

Question 13

How do you feel we're going to do in the next five or ten years vis-à-vis the Japanese? You get an opportunity to visit with some of the top executives in the corporate world. Do you think that the concept of quality is catching on fast enough?

Answer

Let me answer the last part first. The concept of quality is catching on, but the implementation is not going like it should. There is too much emphasis on the ineffective aspects of quality improvement. Many people still think it is a technical problem, not a people problem. The conventional approaches to quality still hold very strong. It is really a lot harder to get straight than people think.

As to whether we will catch the Japanese or not, that varies by industry and by individual organization. It is only a matter of working hard and understanding one's market. There are many American companies who have outpaced their competition. But frankly, American management does not work hard on prevention. They will strive until they are exhausted to overcome some problem once it exists, but prevention is thought to be for sissies.

The areas where the Japanese excel are in high production, which was invented in America. Our policy has been to make 1200 in order to have 900 that we could sell. They learned to make 1200 to have 1200 to sell. And because of their determination to meet the requirements, they have a different approach to things. If a radio plant, say, is set up to produce 100,000 sets a week, they learn how to do each little job and work like crazy to improve continually. Defects are considered catastrophic.

When the marketing department comes and says they can only sell 60,000 radios a week, they don't cut back like the Americans would; they get a new marketing department that can sell 100,000 a week. The Japanese get their priorities straight.

On the other side of that, consider the case of one American company that is in the floor covering business. One area of its market was suddenly attacked by the Taiwanese who appeared with a product that was just as good and cost 60 percent as much. This company immediately cut its prices to match the intruder and then set about learning how to produce at that rate. That group got really serious about quality, and it wasn't too long before they were profitable again. Now they have gone back through the whole corporation in that manner and really present a formidable foe to anyone who wants to compete with them.

Most U.S. companies have the resources to whip their overseas competition. They just need to take quality seriously. But being serious means being fanatical [see Question 51]. You all may have seen the "Fanatic Cards," so you have some idea what has to happen.

Question 14

Conformance to standards seems to be a way of life for the Japanese. When their young people go to work for a company, the idea is already instilled in them. U.S. industries are rapidly losing ground because a lot of developing nations are producing better quality products. Do you get input from the educational field regarding efforts to train our own young people, tomorrow's managers, in this way of thinking before they enter the work force rather than after? What is the quickest way to turn this trend around and keep our industries competitive?

Answer

The quickest way to turn around is for the management of companies to take charge of the education of their people and help them to learn the necessary things. The Japanese don't come as programmed robots. Remember that before the war, the worldwide definition of tacky and unreliable was "made in Japan." They changed because they had to change.

I have a lot of contact with the educational field in terms of business schools and such. However, they are not doing much about quality, at least in the terms we would like. They still treat it according to conventional wisdom. We provide those who ask with case histories, films, and enough material to supply good, viable information to the students. The Fuqua MBA school at Duke, Crummer at Rollin, Harvard, and several others have taken advantage of this material. I agree that it is important. As it is, business schools are actually creating business for consulting firms by teaching quality wrong.

When I speak to business classes, the young people are right up to date, but the professors are singing the wrong tune—not all of them certainly, but enough. However, I would not expect schools at any level to be responsible for teaching ethics or personal values. They can support and reinforce the ideas, but the values that underlie our message and methods should come from the home.

Japanese companies probably do more education and training than their schools. I think that is an important part of a company's operating policy. I was in charge of hundreds of people and millions of dollars before ever being offered a course in management. ITT, however, did a great job of executive education back when I was there.

When the corporations in this country tell the schools that they are not satisfied with the way quality and other things are being taught, things will change. It is silly to hire college graduates who can't write a decent letter and who don't know if accepting a shopping bag full of small bills would be proper. But we, the taxpayers and employers, are the customers of the school system. They think they are doing what we desire. We need to let them know we are unhappy. Changing the national process is going to take a while, but demanding that it change is always the way to start. Companies, even whole industries, can change rapidly. When the quality improvement process is installed and operating, everyone gets right with it. It has to become part of the woodwork. That is why the education part has to go on forever. As I have noted, we have set up a complete college for some companies so they can teach and continue the quality process on their own. It works very well.

Question 15 _____

There is a lot of talk comparing U.S. and Japanese management which glorifies the Japanese style. Do you think Japanese management techniques are superior to those of the United States? If so, what is the major factor that makes them more efficient or more successful as managers?

Answer

I think U.S. managers are the best in the world when they work at it, which isn't too often. The Japanese in the management field are intense about their jobs. U.S. managers, with a few notable exceptions, don't give the job their full, undivided attention day and night. They do not look for ways to finalize problems and opportunities forever. The Japanese excel at repetitive operations like mass production because they will continue to study something and work on it even after it has been squeezed for every drop of progress. They always find one more drop.

When Sony set up a plant in San Diego, which employs U.S. workers exclusively and is the most efficient one in the Sony system, they made the access holes in the TV chassis larger because Americans have bigger fingers than Japanese workers. Would we have done that?

When I showed a Japanese engineer a press shop operation where a worker was feeding a machine one piece at a time, each time pulling his arms out of the way, the engineer looked at it and asked, "What does he do while he waits?"

These stories illustrate the Japanese flair for detail, which is both good and bad news as far as management goes. It is why they are successful. It is also why they have problems dealing with abstracts. But they are creative and relentless. Business comes first, second, and third in their lives. A Japanese manager who comes right home from work instead of stopping off for a few drinks with the boys loses face with his wife. She will think he has no future. Go home and try that.

PCA has a licensing arrangement with the Japan Management Association. They teach what we teach and consult in the same way as we do. Their people came here to learn. The reason they are interested in our system of quality management is that they see a new generation of managers taking over in their country. These people believe all that has been written about what makes quality happen over there. They are putting their faith in the quality professionals and in the systems of statistics, quality circles, and such.

However, what made the Japanese do well at quality was the dedication of senior management. They realized that they could not mount a worldwide service department to keep their products operating. They had to build them right the first time. So while Americans were learning how to sort semiconductors into various levels of reliability, the Japanese were learning how to build semiconductors that didn't need sorting because they didn't fail.

Even today, they get twice the yield from the same machines that we use. They do it by understanding the process and by working hard to continually improve it. Our people compare yields, which is silly.

For instance, I am frequently asked why I do not recognize that there are certain "natural laws" that keep everything from being right. Not all sand is the same, not all cotton has identical texture, not all people are the same, and so forth. I realize that the world is not served up to us on a silver platter. Things change all the time.

Everyone who studies about electricity eventually learns that $E = IR$. R stands for resistance. When utilities build power plants, they have to account for the R because it means that a great deal of the power is lost as it runs through the lines. Sometimes as much as half of it disappears between the supplier and the user. Such an obvious item certainly cannot be ignored. Whole sections of the utility staffs do nothing but deal with R.

Now it is being learned that using certain materials and certain temperatures can make resistance disappear. $E = IR$ suddenly becomes $E = I$. What you send is what you get. This is something that was not known before. If scientists had taken resistance as something that could never have been eliminated, then it would have been with us forever, which would prove that it could never be eliminated.

U.S. managers are not very open to new ideas. I have toured this country from one end to the other for years trying to get people to look at quality in a different way. Hardly anyone would even think about it. They KNEW; they didn't have to question. Everything must be questioned continually. I learn every day. I would rather have U.S. managers than any other, but they need to learn how to get things done, and then do some more.

Cultural differences are a fascinating aspect of business. When ITT established quality councils around the world, I had to deal with many different nationalities. It was fun to learn what to do and what not to do. The normal courtesies differ from nation to nation. Here, taking flowers to a hostess is considered a gesture of politeness. In parts of Europe it is not considered so, particularly if they are carnations, which are associated with death.

Georges Borel, the French ITT quality executive, was the chairman of our European executive quality council for several years. He was also my personal guide on understanding Europeans. He used to tell me the difference between heaven and hell.

"Heaven," he would say, "is a place where the police are English; the chefs are French; the mechanics are German; the administrators are Swiss; and the lovers are Italian."

"Hell," he would continue, "is a place where the chefs are the English; the mechanics are French; the police are German; the administrators are Italian; and the lovers are Swiss."

Obviously, entire nations cannot be characterized accurately, but it does show how people look at themselves. The Japanese seem to see themselves as silkworms that have to produce the same kind of silk all the time, to be consistent, to not stray from standards.

This is hard for the western world to understand because individualism, being different, is what we are all about. I think individualism will prevail in the end.

So when trying to teach in other cultures, it is preferable to train nationals to do the teaching. The material is translated with their participation in order to pick up the subtleties of the language, and they conduct the discussions in accordance with the practices natural to their national culture. Americans who particpate in any of these classes have to learn to modify their approach because it is not considered proper to be quite as open as we are used to being.

Question 16 _____

Suddenly in England we're beginning to see keen competition from the Pacific rim nations. What's happening? Are these people fanatics? Do they operate by another creed?

Answer

I have talked a lot about Japan, which is hardly the South Pacific, so let me comment on Hong Kong, Malaysia, Taiwan, and a few other places—the ones I assume you mean.

Yes, they are fanatics. They will do whatever it takes to invade and conquer a market. They do not have much of a home market because most of the citizens in these areas are relatively poor. One of the advantages companies have over there is the low cost of labor—paying people just a few dollars a day for their work. If the processes can be designed so that they are less demanding, then the labor content is quite low, well below what has to be paid in England.

However, that is only one small part of their competitive edge. The Chinese, whom you will find behind most of these companies, are marvelous business people. They are frugal, they prize cunning, they work like crazy, and they thirst for your customers much more than you do. They will take on any task, work any number of hours, and sacrifice whatever it takes in order to overcome. They are well-financed and tough.

We in the western hemisphere owe a great debt to the leaders of mainland China over the past years for holding down its progress. If that nation of a billion business people ever gets turned loose to use their natural talents, it will be a tougher world in which to earn a living. That doesn't appear likely for some time. There is still a great deal of discussion about China becoming open, and my observation is that there is a long way for them to go even if it is possible politically.

There is no infrastructure in China, for instance. Services—roads between towns, mail service, power, the telephone system, bank systems, water, health care—are very basic. All of that can be overcome with money and energy, but it is going to take a while. Thus we're given a reprieve, temporarily.

The future can be seen in Hong Kong. There the business people are insatiable. They have factories and banks side by side. They think of business all the time. Calm, rational westerners spend a day there and come down with "Hong Kong fever," which results in spending more than planned to buy things that were not considered necessities until that moment. Bustling on the Star Ferry, which runs precisely on time back and forth across the harbor without turning around, they buy until they are exhausted. Then they stop at a shop in the hotel lobby for just one more. It is like being an alcoholic.

There is no reason why Britain cannot become the Japan or the "North Sea rim" of the world before the future I'm describing arrives. Here is a land much closer to its market. Sixty percent of the resources in the world are controlled by Britain's natural market of Europe and the United States. There is a large work force available, much financing, and plenty of transportation. Wage levels are higher, of course, but they make up a small part of the total cost of production.

All that is necessary is that those who run Britain's companies and those who run the unions get mad about it all. Why should other nations come in and take markets from the English? They are able to because of sloppy work and lack of attention to the customer. This is why our quality improvement process has been so successful in the United Kingdom.

Organizations are finding that they can compete and win. It is not necessary to have barricades in order to survive. There are many examples.

People get ideas about how things have to be and they don't want to change. I had always thought that acupuncture was quackery. The last time I was in China, I had a sore back from sightseeing and car riding. I went down to the spa in the hotel, only to find it was not what I thought it was. They sent me over to see this little fellow in a white jacket and after we sign-languaged a little bit, he understood that my lower back was hurting.

He got out a statue of a person with little black dots all over it and pointed to a set of them around its upper back and shoulder. Then he lifted a black case out of the drawer and opened it to display about fifty needles of various sizes. He smiled and gestured for me to lie down on the table.

I declined. But at that exact moment my back stopped hurting and didn't bother me for the rest of the trip. And they say that acupuncture is not effective.

Question 17 _____

How do you sustain the momentum of the quality improvement process when the bottom line starts to slip?

Answer

Momentum is not what keeps the quality process moving or makes it a permanent part of a company's structure. Necessity and success push it along. Once people learn to work this way, they will not want to give it up.

If something is worth doing, it will continue regardless of temporary situations or setbacks. And the best way to keep the bottom line attractive is to learn to prevent loss and waste.

When I started in the quality consulting business, I thought the clients most likely to be attracted would be those who were losing money or were in terrible trouble. But it didn't work that way. They were instead well-managed, profitable companies. And when the recession of 1982 hit, they stayed right on with what they were doing. These companies realized they were in business for the long run. Their profits took a hit at that time because the whole world of business hit an air pocket and dropped about 5000 feet. But they stayed with it.

Our company fell into the same pocket. For a couple of months the flow of new clients dried up while everyone waited to see if the recession was here to stay. The pickup started right after that and we had the same type of companies calling. We tried to learn not to let expenses get ahead of revenue, but that is hard to do when you're growing.

So the momentum is built into management's confidence in the company and in its confidence in the concepts of quality management. Good managers, whether personal or professional, do not change policy or objectives because of something that jumps up in front of them. Strategy changes are continual; policy ones are not. We wouldn't think of running out and robbing a bank to meet the payroll, but an across-the-board pay cut might be considered in a pinch.

Process industries, including insurance, textiles, steel, semiconductors, and many others have yields that are really terrible. They especially need to concentrate on learning how to get things done right the first time as the most rapid way to reduce expenses and increase profitability. They also need to concentrate on much more output with the same number or fewer employees. That is what needs work all the time, but particularly in times of downturn.

Question 18 _____

It's a regrettable facet of human nature that we tend to learn more from our failures and disasters than we do from our successes. Your own efforts at losing weight and stopping smoking

were in the face of disaster. Have you had any failures with the
Crosby medicine you prescribe, and what causes, if any, lie at
the root of these failures?

Answer

The whole system for quality improvement I refer to as the
Crosby Complete Quality Management System was born
from a long series of "inadequate successes," if not necessarily
failures. To me, when I was trying to convert the companies
I worked for, their suppliers, and the rest of the world to a new
way of thinking, any move forward, however slight, was a success. I was like the insurance salesperson who, when thrown
bodily out of the prospect's home with the dogs chasing and
snapping at his heels, marked it down as a "possible."

The lack of response by senior management over the years
I considered to be because I had not found a way of explaining the concepts in an understandable way. The general lack
of progress in the quality control field I attribute to several key
thought leaders who have never been able to understand what
I am talking and writing about. This is primarily because they
have never read it or heard it.

At this time most of the country is still in darkness, fighting
the quality battle the old-fashioned way, with excuses. I consider the fact that these people have not seen the light a failure, and one that, as you said, drives us ahead to find new
ways.

On the other hand, I don't know of anyone who has seriously applied the quality improvement process as I have described it who has not been very successful. Everyone got better. Some have made incredible achievements. Some have
succeeded only wonderfully. The system works and has been
proven beyond the shadow of any possible doubt, and it is
continually being improved.

Where failure happens is when an organization takes the
job on without the commitment of top management. We

did have some early failures and realized that we had not insisted that senior management come to class and really understand what was going on. We had failures when the new client came only because a customer was beating them up on quality.

Once we realized this and learned to qualify the companies before starting to do business, we never had any problems along that line. But we have had to turn down some nice agreements in order to stick to this policy. One was for 6 million dollars. There was a company that wanted us to change the college dramatically and would not deliver their management, would not even talk to them about it. So far, they have spent a lot more than $6 million making their own internal material and have gotten nowhere. It is a shame. But people get caught up in misunderstandings and in their own need to drive the wagons over the cliff personally.

We see companies where the top management is out preaching about quality under the impression that everyone understands it in the same way. They think that back in the factory they are all doing what is being talked about up on the podium. Because there is no serious definition or direction, no one knows whether there is success or failure.

Quality improvement is a complex business.

Quality Thinking: Toward a Mature Philosophy of Quality

Question 19 _____

What do you think is the most important thought in the philosophy of you and your organization?

Answer

The thought that will do you the most good is "prevention for the purpose of *causing* defect-free work." That will reverse the way a company normally operates.

Neither your company, nor any other company for that matter, has ever put out an advertisement that says the products or services produced will contain errors and defects. There is no discussion of such things with customers. Service operations and field support are set up in the unlikely case that something goes wrong.

Yet the company is run to accommodate nonconformances on a regular basis. Waivers, deviations, and such are so normal that there are actually procedures and training to cover them. Nothing is ever delivered exactly like the requirements state. It becomes a tradition and it has to be changed.

It is hard to realize how deeply such practices can become ingrained. It is the ultimate in self-fulfilling prophecies. A few weeks ago I was privileged to be the luncheon speaker at a Department of Defense quality meeting. There were about 800 people in attendance, mostly senior purchasing, quality, and contractor people. There were several three-star officers and many corporate presidents.

I estimated that their combined experience had to cover 25,000 years, at least in the quality business. Because of that, I said I would like to take a poll. I asked for a show of hands on how many of them knew of any product or service that had ever been delivered to the government without a waiver or deviation. Did they know of any? No hands, not one. A lot of embarrassed looks, but no hands. That is one good reason why quality is so bad nationwide.

All of that could be rooted out of the DOD in a few years with some new education and direction, just like it can be rooted out of any company. There is no reason for not taking requirements seriously, and people will do that when it is what the organization believes.

So if you can take away an understanding that it is very much to the benefit of stockholders, employees, customers, and suppliers to begin a tradition of doing things right the first time, this will have been a successful experience for you *and* for them. Your company will be much more profitable, will have a higher chance of eternal success, and you can lead a hassle-free existence.

Get clear on what zero defects means, and keep others aimed at it as the only goal worthy of the company.

Question 20 _____

I have a feeling that when I get home after this course, people are going to ask me how this approach to quality improvement differs from other approaches we've tried. If you approach anything with enthusiasm and you go through all levels of the organization, you'll get results in the short term. What makes this process unique?

Answer

Enthusiasm is a wonderful thing and lessons are very helpful, but they can only take you so far. I have a friend who teaches motivation. He is well known throughout the country and is a wonderful person.

Once in a while, we cross paths and have a few moments together. Just lunching with him gets me so pumped up that I can't sleep for a couple of days. The last time he was here he asked me to tell him my biggest personal problem—not something like failing to get on *Fortune's* billionaire list, but a real problem.

After some thought I told him about the 17th hole at the Bay Hill club. It is a 190-yard, par-three water hole with a skinny green, and it drives me nuts. I have never hit the green yet. Knocking a ball 190 yards is no problem for me, but I have never solved this hole. I worry about it while playing all the rest of them.

He leaped right on this telling me that I lacked self confidence and assuring me that I could overcome if I realized that it is only necessary to assert myself. I got excited and we went to my car, drove out to the club and right up to the 17th tee. It is near a residential street. Being a Florida golfer, I always have my sticks in the trunk, so I took out a three iron and an old ball and walked up to the vacant tee.

He shook his head.

"Now you see, Phil, you have an old ball there. That demonstrates a clear lack of confidence." He looked at me in disapproval.

Chagrined, I went back to the truck and picked out a brand new black Titleist and placed it on the tee.

"Now," he said, "Take a couple of practice swings."

As I gave him three of my best, he gazed intently and then smiled at me.

"Is that your regular swing?" he asked.

I nodded in assent.

"Perhaps," he noted sadly, "You should put the old ball back on the tee."

What we are working at here is entirely different from anything else that has been done in your company. The aim is to help you attack a problem that everyone agrees exists. We are talking about quality improvement. Why are you here? Your companies sent you here to learn how to improve quality. Why did they send you? Because the quality is not satisfactory. It may very well be lousy.

We didn't promise you anything. We didn't entice the management to get involved. You all decided on your own. Improving quality requires a culture change, not just a new diet.

Look about you, everyone else is here for the same reason. It is like going to Weight Watchers. Do you see the Miss America candidates there? No, you see overweight people there, people who want to change the way they approach food, who want to learn to understand it and learn to benefit from it.

How did they get overweight in the first place? Because they ignored the subject for years. Why is your quality unsatisfactory? Because no one paid any attention to it until complaints began to roll in from the marketplace.

I'm not talking about improvement sufficient to let you squeeze into the old tuxedo just one more time. I'm talking about learning how never to have another problem with quality. That is something more than running a group of seminars on inventory or some single facet of an organization.

This is the cold, hard reality of business. If management cannot get everyone to understand quality the same way and work at it with sincerity, then it may be all over. That is reality—not a program, not enthusiasm, not something we are going to do because we feel good. It is an essential, but it does not come automatically like breathing. It has to be worked at every day—like golf.

Question 21 _____

In one of our safety programs at my company, we related safety to profit, productivity, and morale. One of the speakers here has said that quality is "first among equals." Where do you place quality in the overall scheme?

Answer

We are the ones who told that speaker that quality is first among equals. Safety is a great analogy for understanding quality. Everything safety is about relates to the absolutes of quality management.

First, we figure out how to keep a safety problem from happening. Then we make up a list of requirements everyone can understand, insist that everyone conform to them, help the process along, and measure the results. The idea is zero accidents, not reduced accidents; not eliminating lost-time accidents, but to eliminate any kind of accident at all.

So quality is part of each of the functions you mentioned because it lays out the system of making them happen. If we make our profit goals, but don't pay our bills, then we have not met our profit goals. If we deliver on time, but the product has defects, we have not delivered on time. If we meet our safety objectives, but damage somebody, we have not met our safety objectives.

We worked with a steel company once whose chairman was not at all interested in helping drive the quality process. He thought people should just do their jobs. I talked to him a great deal to little avail. Then one day his company received a safety award, and he was so proud he could hardly stand it. Safety was his baby. I told him that safety and quality had the same basic structure. His company was first in safety and last in quality because management didn't take quality seriously.

He looked at me and said, "Now I understand." And from then on things turned out right. It also works in reverse. Many executives who are sincerely dedicated to quality of product and paperwork will unthinkingly shortchange safety. They will forget to set a good example by wearing safety glasses in the proper areas, or by being concerned about noise, or by participating in the safety awards presentation. I used to get a lot of free lunches by betting general managers that I could find someone violating a safety rule during our tour of their facility. It was always the general manager.

Question 22 _____

I am really interested in the Four Absolutes. They contradict what we might normally believe. Could you say something about their origin and how they developed?

Answer

That's a good question because these things didn't just fall off the turnip truck, as we say in West Virginia. For those of you who have used your class time to figure your income tax, the Absolutes are the basic conceptual foundation of the quality management philosophy I have developed over the last 35 years. As you may know, they are:

- Quality is defined as conformance to requirements, not goodness.

- Quality is achieved through prevention, not appraisal.

- The quality performance standard is zero defects, not acceptable quality levels.

- Quality is measured by the price of nonconformance, not indexes.

When I started as a junior technician in 1952, I had no industrial experience. So I studied hard and learned the quality control business, which taught me that nothing could ever be right all the time so we had to learn how to contain errors as much as possible.

Then I learned about reliability, which offered even more proof of the inevitability of trial and tribulation. Both disciplines—quality control and reliability—provided convincing statistical proof of this rather dismal picture. But I began to question why it had to be like that, and the answer was always that the laws of probability dictate that sort of performance, and besides, management wants it that way.

I couldn't believe that managers never wanted to get any better. They were always talking about success. Yet the "proof" was all about. Nobody delivered anything defect-free to anyone.

When I got far enough up in the organization at Martin and ITT to understand what management was about, I began to realize that they did not understand quality any more than we quality control people did. They had no idea that it could be managed. They felt it was all a matter of chance. Containment was the order of the day. *Causing* quality never occurred to anyone.

So to try to make it clear to management, I began explaining that it was up to them to set clear requirements and then help people learn how to meet them. I told them that the thought of error being inevitable was a self-fulfilling prophecy. If you think it has to be that way, it will be that way. Management people understood this; hence the Absolutes began to develop with the thought of management stating clear requirements, helping people meet them, and insisting that requirements be taken seriously instead of being waived all the time.

However, the quality control and reliability people, particularly the "gurus" of the field, denounced me as impractical and unscientific. Most of the denunciation has subsided, but there is still some, which is okay with me. As your question points out, this way of thinking was—and still is in many cases—a contradiction of conventional wisdom. It is hard to get people to change their minds once they are convinced of something.

The Absolutes, as they now stand, were laid out as a teaching aid for the first Quality College course. I didn't formalize them into their current format until early 1980. There had been five Absolutes originally—as I had developed them over the years. They were listed in *Quality is Free*. These had been the ones used inside ITT. Three of them—the definition, measurement, and performance standards—I have kept the same. I eliminated one that said "There is no economics of quality" (although there isn't) and another stating "There is no such thing as a quality problem." I replaced them with "The system of causing quality is prevention, not appraisal."

The principles about economics and quality problems were designed for quality professionals. Therefore, they did not mean much to managers. They are both still true, but we teach them in different ways. The "economics of quality" is some mysterious calculation used to show that it costs too much to do things right. It is based on the same erroneous assumption of the inevitability of error. I wanted to get people to stop using the expression "quality problem." This makes it sound as if the quality department is responsible for fixing the problem. I estimate that from the time I first began to think such heretical thoughts until things reached their present status was about 20 years.

If I hadn't been a hands-on, real-life type of worker, the Four Absolutes would probably never have evolved. I have to be under stress in order to be creative. Perhaps one account of some of that stress will illustrate the learning and development process as it happened in my life.

When I was quality manager of the Pershing Weapon System at Martin in the late fifties, our customer technical director was the Army Ballistic Missile Agency (ABMA), which was headed by Dr. Wernher Von Braun, the German missile expert. They were very serious about quality and were not too thrilled with the techniques used in the aerospace industry. They had a hard time understanding the concept of material review, in which requirements could be changed in order to meet some nonconformance problem.

One part of the construction operation that caused a lot of trouble was soldering. In every electronic system soldered, connections were a primary source of unreliability. It used to drive me nuts because there was an elaborate system of industry standards that permitted soldering to be nonconforming.

These were standards that everyone considered to be quite rigid. One "no solder" per 10,000 joints was allowable. "Rosin joints," "crystal joints," insufficient solder, and all the other exceptions to conformity were neatly described, along with the number of times they could be expected to occur. Everyone had color photos of what bad—but acceptable— joints looked like. As a result, soldering and other connections were just awful, everywhere, and no one thought they could be better.

ABMA developed a solder procedure called PSD1 which they wanted us to use. It stressed a "J" wrap rather than the traditional one and a half wraps, and used minimum solder rather than a glob. They made it very clear that if we did not want to do it that way, the contract would be in jeopardy.

I happily agreed and we set about it on the condition that there were no acceptable nonconformances. Everything had to be done correctly. This was the first time I had ever insisted on such a thing. To make this happen, our folks had to learn how to solder to this specification. A school was created inside Martin to teach everyone how to do it that way. All the suppliers had to comply, and they did. We taught their teachers.

Every solderer in the system had a number, and any time problems were found the person responsible was retrained if that was the problem. In a short period of time we learned how to solder that way, and during the lifetime of the Pershing, solder was a completely insignificant problem.

We had learned how to prevent something from happening. Unfortunately, I had a great deal of difficulty convincing many other people that this could be done in other areas of the business. It takes a while to overcome deeply rooted beliefs.

A story from my days as a hospital corpsman in the Navy will prove my point. I would hold sick call every day, and one day this marine walked in. When I asked him what his problem was, he told me he was dead.

"You don't look very dead to me," I said.

"The sergeant is having trouble with it too," he replied. "But actually I have been dead for some time."

I called the doctor out to examine him, and the doctor explained the way the blood system works.

"Look at this drawing," he said. "The heart beats and the blood goes around the body. Now when the heart stops beating, as it does when a person is dead, the blood doesn't go around any more. So dead men do not bleed."

"I see," said the marine.

The doctor told him to take the drawing back to the barracks with him and keep repeating "Dead men do not bleed" over and over.

The next morning the marine appeared as planned and I asked him what he learned.

"Dead men do not bleed," he replied.

The doctor was pleased with that response and then took a scalpel and nicked the man's finger. Blood poured out. The marine looked in amazement.

"I'll be darned," he said. "Dead men do bleed."

See, that's how hard it is to change a person's preconceived, hard-wired notions. Almost everything that is believed categorically is not correct. Columbus knew the world was round but thought it was only about 10,000 miles in circumference. Others knew it was 25,000 miles and that you couldn't carry enough food and water in those little boats to make it across. Columbus was out of about everything and 30 percent of the way to China when he found the new world.

My high school football coach believed that the forward pass had no future. "Only three things can happen," he would say, "and two of them are bad." Parental advice is usually out of date, too. I could go on.

So when I began to question quality and wonder why things were not being made correctly, I started to look at what the word meant. And here's what I noticed. All of the articles in the quality magazines and the books by the business school professors talk about quality as though it were something everyone understands completely—and in the same way. Notice how everyone speaks knowingly about "excellence." You tell me what that is so I can understand it.

When I began speaking about quality being conformance to requirements, and zero defects being the performance standard, the more experienced people would take me aside for a big brother talk. I would never get any respect if I didn't tune in to reality, they would intimate. Even today, people insist on misunderstanding. Two recent books quote me as saying "Quality is conformance to specifications." Then they go on to say how shallow this is and talk about customer requirements and such. When we write them a courteous letter pointing out that they had read my five books on the subject wrong, they write back a courteous letter apologizing and promising to change it in the next edition. But they never do.

One other incident sticks with me. At my very first reliability conference, back in 1956, I came across a man sitting in the lobby of the Statler Hotel in Washington, D.C. We started to chat and he showed me the contents of a box he had with him. Inside was a machine. It consisted of 100 discs, each about the size of a 45 rpm record, mounted on a bar in parallel with each other. They looked like a giant submarine sandwich. Each disc had a little red mark in it, and there was a wire stretched across the front of it.

"Each of these red marks represents one percent of the circumference of the disc. So there are 100 discs and each one is 99 percent reliable. When I spin them, it is like a system with 100 components each of which is 99 percent reliable; or like a paperwork system, such as the payroll, where each of 100 procedures is performed right 99 out of 100 times."

"Now given such a situation, what is the reliability of the system?"

"Must be about 90 percent," I said.

"Aha!" he replied. "It is really 36.4 percent because you must multiply each 99 times the other. Ninety-nine times 99 times 99 and so forth for 100 times."

"That is why nothing works!" I said.

He shook my hand. "Now you understand reliability," he smiled.

When I got back to my company, I went on a campaign to get rid of "acceptable" quality levels. People asked me if I didn't have anything else to do.

So anyway, that is a little bit of how the Absolutes came about.

Question 23 _____

Aside from your work in the quality field, you are an entrepreneur. What are your views of entrepreneurialism and of your own experiences as an entrepreneur?

Answer

That is a really funny thing. I get asked to make talks to college groups who are studying entrepreneurship and I never considered myself one of those. All during my career I thought I was just a good team player moving along with the flow. But now I have reached an understanding about it and can realize why things worked and didn't work over the years.

A person is either an entrepreneur or not. That person is born with the intent and learns how to implement—if he or she can recognize this entrepreneurial talent. It is not always easy to know. Some people die never having recognized that they had this characteristic, because no one ever told them. As I look back, I can see that mine was wired in, like a printed circuit board, even though neither I nor anyone else realized it was there.

Now I never was a kid who had a lemonade stand or a paper route. I looked at those things, which fathers think are a wonderful idea, as a lot of work for very little return. I was considered someone with no serious ambition who would rather read than do something useful.

I invented my own jobs, like gathering up the shopping carts in the lot of the supermarket or pushing the groceries out to cars for ladies who gave me tips. I was always a pleasant kid and they thought I was kind to do it. Today it is routine, but at that time it was looked on as not exactly legitimate work because it involved little discomfort and no regular hours. My peers, who braved freezing weather to deliver the *Wheeling Intelligencer* while I was snug in bed were considered superior beings to me. I know my father was always secretly disappointed that I did not have more ambition. In the years I worked for large companies I found the same attitude. If you weren't at your desk, sleeves rolled up and surrounded by paper and old coffee cups, you weren't working. My style, which is based on thinking about things wherever I happen to be, from walking around the block to playing golf, was not readily understood.

However, I learned how to get results, get them quickly, and at the same time have everyone happy it was working out so well. This was something that took some doing because I had been a follower more than a leader throughout my life.

I went to college to study podiatry because that was what my family wanted and I couldn't think of anything else. That was after World War II. I went through school sort of sleepily and graduated just in time to be recalled for the Korean war. I had always liked school but never cared for doing the assignments or studying what the class was working on. I got straight Cs all the way through from the first grade. But I read a couple of books a week and in doing so picked up a lot of information.

When I returned from the Korean fracus, I decided that I should take charge of my own life. I went to the state of Ohio employment office and wound up working in the quality department of Crosley, in Richmond, Indiana, as a junior technician.

It didn't take long to learn that most people really don't know much about what is going on except in their own specific area, and that no one knows what anyone else is doing. So I began to apply some imagination to the job and the next thing I knew 13 years had gone by and I was in New York working for Harold Geneen as a vice president of ITT.

What had made this progress possible was not my technical skills or my financial knowledge. It was the realization that the traditional concepts of quality management were ineffective. They are aimed at a "factory" environment and are primarily technically oriented. Management didn't have the faintest idea what was happening, and the quality people put them to sleep trying to explain it. I began to develop my own ideas of what quality improvement was all about. My understanding that quality is conformance to requirements, for instance, came early in my career, when I realized that the company's reputation and future rested entirely on my opinion of whether something was right or wrong. The other side was taken by the manufacturing people. We argued continually. We didn't pay too much attention to what was actually supposed to be included in the product. If we could agree, out it went.

When I brought all the "requirements" together and we agreed on what the product was supposed to look like, then the arguments went away. Opinion did not count any more. That may not sound like an earth-shattering development, but it was a whole new look at things. At the same time I began to realize that the shop was the victim of the system, not the perpetrator. It also began to dawn on me that prevention was something that could actually be done.

However, when it came time to communicate these ideas, I was not too adept at it, so I went to Toastmasters to learn to speak and I wrote tons of stuff in order to learn the craft of writing. I also discovered that I was a fairly creative entrepreneur. I could take a situation, grasp the essence of it, and then turn it into a success. I am very people-oriented and never see them as a threat, only as a help. Thus there are people who disagree with my philosophy but who like me personally. There are also some people who think I am flakey.

An entrepreneur, then, is anyone who can develop a plan, explain it to others, get them to eagerly help, and at the same time keep a consistent pressure on the enterprise—that is the kind of person we are talking about. So the entrepreneur is one for his or her entire lifetime, inside a company or outside. I wrote a book called *The Art Of Getting Your Own Sweet Way* which tells how to do it, even though I didn't realize it at the time.

Every business school has courses now on how to become an entrepreneur, and many corporations have developed internal programs in order to encourage people to develop entrepreneurial skills. However, it is not possible to select someone and send that person to entrepreneur class and expect some magical transformation. It isn't like learning how to play the piano.

If a course is set up and is made difficult to attend, those who should be in it will find a way to get there. Management always picks the wrong people. It's like the teacher picking hall monitor. I would never have picked myself to be one, and neither did anyone else along the way.

Entrepreneurs should not be seen as flagrant risk takers. They are very careful to think things out properly. The risks they take are in entering uncharted areas, not in doing so in a leaky boat. They are usually well prepared, have their eyes wide open, and know what to do when they arrive somewhere. The difference between an entrepreneur and an executive is that one is customer-oriented and one is management-oriented. The executive runs what the entrepreneur brings together.

I have been concentrating for the past three years on developing a young management team for this business. And now that team runs the company completely. My commitments now are more toward teaching and marketing, helping with strategy, and my writing. But they run the company. I had very little problem in giving up that authority because I do not consider myself a businessperson. I am more a pragmatic philosophical type. If you're interested in more details about this and related issues, it is explained in *Running Things*. This was an essay I started which wound up as a book.

Question 24 _____

First, the quality improvement process as you espouse it is attended by much bounce and enthusiasm, but the language in many cases seems negative. For example, you use words like prevention, elimination, vaccine, conformity. How do you explain this apparent ambiguity?

Second, regarding conformity, most of our outfits are not after conformity so much as they're seeking to stand out, to be different, to be unique, to get ahead. That seems, in my mind, to involve nonconformity and risk. How do we move beyond conformity?

Answer

There are a couple of different things involved here. I'll answer your questions in reverse order. First, we are not advocating conformity in the sense of being like everyone else. We are discussing requirements we establish ourselves, and taking the necessary actions to complete them. Requirements are established by those who run the business. They lay out the specific things that need to happen in order to stand out, to be different, to be unique, to get ahead. The strategy and the function of the organization is described in these requirements. Having spent the time and energy to determine them, it is necessary to be serious about them.

Second, the reason you see bounce and enthusiasm in our operation is that the associates of PCA know what the requirements of the organization are. They know their responsiblity for accomplishing them, and they are recognized for doing so. They also know that when they have problems, they can get counseling or training as needed. But what determines everything that happens are the requirements we have agreed upon.

Negativeness? Some of the terminology you have heard may seem defensive, but I don't see that as negative. In fact, I see prevention as a positive effort. If you saw a sign that said *Last gas for 200 miles*, you would tend to think in terms of prevention and you would, accordingly, purchase at least 200 miles worth of fuel. You mentioned the word vaccine. Children are now vaccinated against diseases that were common during my childhood and perhaps yours. We insist that it be done. Again, that is a positive thing.

Question 25 _____

Conformance to requirements means that the requirements must be tangible. Should the majority all be in writing or just the most important ones?

Answer

It depends on what is being done. Requirements are answers to questions and the agreements that result from those answers. When I go skiing, I always take a lesson the first day and the instructor gives me requirements. Those are not written down right there. It would be hard to read something while moving down a slope. There are books and films that contain exactly what the instructor teaches. Out on the slopes, there are requirements that define the skill levels of each slope, the dangerous areas, where meals are, and the like. Those requirements are written on signs and coded so as to be "idiot proof." It is a bad mistake to misread one of those signs.

Another example: My wife and I bought a lot, tore down an older home, and are building a new one. Many, many requirements are involved. Hundreds are written down, some are contained in legal documents, some are handshakes, some are assumed. We are doing what it takes to be protected by the legal system, to pay bills on time, to construct according to agreement, and to provide the opportunity for something to be permitted to occur that was not foreseen.

Similarly, I have set up trusts for my children and for charity that are supported by the revenues from my writing. These are administered by third parties and function according to specific requirements.

There are all kinds of requirements. I have unwritten agreements with my grandchildren, two of whom do not read yet, that I will do whatever it is they want to do, whenever they want to do it. So if a small voice calls the office and asks for me, I am pulled out of whatever I am doing.

So we need to keep "requirements" in proper perspective. It is we who are the masters, not the requirements. They serve to mark an agreement between people and should take whatever form is necessary. They must be respected and never altered except by agreement between those who created them.

Question 26 _____

I'm with atomic energy and we do very technical things, as you can imagine. The concept of requirements has a special meaning in our organization, and your concept of requirements is a little different. Could you say a few words for those of us who work in new industries or fields in which the requirements are still being established, and whose job it is to find out what the requirements are and to decide what to do?

Answer

Actually the concepts are identical. Remember that requirements are the answers to questions. Sometimes it takes a lot of effort to lay them out. If the job description is of one who creates requirements, then the idea is for them to produce something that other people can understand and implement. There should be no requirement that cannot be met. Sometimes we may not know everything there is to know, but we have to begin and learn as we go forward.

There is a parallel in atomic energy work. Many unknowns exist and in order to get enough information to define or clarify new requirements it is necessary to enter unknown environments or positions. To do that properly, it is necessary to define experiments and investigations. All of these must be performed as agreed, according to the requirements, or it is not possible to know how to deal with the results. Once the situation is understood, then more requirements are created so those who have to make things happen can know as much about the task as the one who investigated it.

So there are a bunch of requirements in everything we do. Some are everyday requirements; some are not clear. We have to be able to count on each other doing what we have agreed to do. As we learn more and improve the explanation of the requirements, those requirements will become more effective.

Everything that happens in any operation today happens because of requirements that exist. They may be written down, they may be verbal, they may be traditional, they may have just been spoken a moment ago. Don't think of requirements as something in capital letters with a specific format behind them. They are just the answers to questions.

Question 27

Is there a conflict between zero defects, doing things right the first time, and risk taking?

Answer

Zero defects is a symbolic way of saying "do it right the first time," for which some people substitute the acronym

DIRTFT. Since this is hard to pronounce, you might try DIRTFOOT, but you'll end up with black footprints all over the place.

I think people get confused when they talk about risk taking. The real risk takers, like those who climb mountains or go into space, make sure that things are right the first time, and they do it before going. They get the base camps set up at just the right place with just the right equipment, or they proof test the engines and train the crew until everything works as planned.

When I left my nice warm job at ITT to go it on my own, people said I was taking a risk. Certainly the decision involved a great deal of financial exposure in that I was giving up an income. But I had saved some money and my plan was to grow the new business on its cash flow.

In order to make certain there was a need—and to be sure I had clients—I wrote *Quality Is Free* and aimed it at executives. If it was well received, I would go ahead with my plan. As things turned out, it was very well accepted. By the time I went out into the world, a lot of people already knew me and that made things easier.

My early clients, like IBM, Tennant, Milliken, Bendix, Celanese, J.P. Stevens, Brown and Root, and others were patient with me. I knew the quality business. They helped me learn the consulting business. It was easy to walk into an organization and say, "Do this; don't do that." The results would always be good because I have seen just about every situation that could exist. However, I didn't want to do that for them. I wanted to put them in the quality management business on a permanent basis. They needed to learn how to run it themselves so they could manage the company better.

The key to "risk taking" is taking time to lay out the requirements, to get clear on everything that is known. Some things are not known and have to be planned for, but if the rest of an operation is well based, the unknowns become something that can be dealt with. When people are too lazy or preoccupied to work out the requirements, then they are taking risks. It turns out, for instance, that several Arctic expeditions never had a chance of getting back because they had not brought enough provisions with them or they had gone at the wrong time of year. Requirements need thought.

If an organization is really to be run properly, everyone has to understand the purpose of it. Everyone has to know the charter of the organization, and they have to understand their personal role in making it all happen. Those are the systems I wrote about in *Running Things*.

The Navy destroyer captain who wants to get close enough to shore to provide relief to a group of trapped marines takes a careful look at the chart in order to know where to place his ship. There is no great advantage in having a combat vehicle piled up on the rocks and out of action. It is a risky operation in that one never knows what the enemy will do, but one can at least know the sea and the resources available. The captain has a solid base of support, a base of well-trained sailors with equipment that has been maintained and is in fighting shape.

This captain knows exactly what the ship can do. He gets as much information on the enemy as possible, informs the crew as to what is going to happen, and in they go. Now a risk would be for him to say, "Crosby, go fire the five-incher." Crosby doesn't know how to fire the five-incher.

Zero defects is doing *what* we agreed to do *when* we agreed to do it. It means clear requirements, training, a positive attitude, and a plan. The conventional notion of risk taking is that it involves leaping off into the unknown. People who leap off into the unknown disappear into that same unknown.

When corporate executives talk about risk taking, they usually mean leaping off into the unknown without fear of punishment should things not work out. They should join the CIA.

Question 28 _____

Is zero defects a goal or a reality?

Answer

Both. It is clearly a goal, but zero defects is very much a reality too. All kinds of things come off as planned. We just don't always recognize it. For instance, all of you went out back of the building today for lunch, the classes broke on time, the tent was up, the tables were ready, the food was hot and tasty, the seats held us as we ate, and we are all back in here right on schedule. The microphone works, the lights are on, each of you has a seat, the "Fanatics" cards are in your hands with each word spelled correctly, and we are all speaking the same language.

That's reality. Zero defects is the result of thinking things out.

The reason the question even comes up is because we have been programmed since our early days to accept as fact that things can't be right all the time. The fact that in real life a lot of things go wrong does not mean it has to be that way. A great many discrepancies occur merely because they are expected to happen. That expectancy becomes so routine that people spend their time learning how to fix rather than prevent.

There is great demand these days for corrective-action and problem-solving courses. There is little demand for prevention-specific courses. That is why it is necessary to think in terms of a whole culture change if we are going to make a difference.

One of the most cynical statements imaginable is known as Murphy's Law: "Anything that can go wrong will go wrong." There is even a followup to that which says, "Murphy was an optimist."

Now following that philosophy makes life a lot easier. If we don't expect much, and don't get much, then there is nothing to complain about. To me that is like not caring if the bowling ball rolls into the ditch or not.

In my days as a reliability engineer, I found that I had stepped into a world where the assumption that failure was inevitable had reached an art form. It was proven, statistically, beyond doubt, that we could not expect our systems to work every time when the switch was thrown. Test results and field performance backed this up absolutely. Nothing performed as planned, so accommodations had to be made. Redundant circuits were routinely included. Continual checking was employed. People sighed continually about the difficulty of the tasks, and the products were very expensive.

As I lived in this situation, I grew more and more frustrated. Because of the unspoken agreement that nothing was going to be right, there was little serious effort to produce reliable components, for instance. Instead of learning to understand the process and eliminate the causes of errors, concentration was on bigger and better testing to find the bad components and keep the good ones. There was a whole class of "high reliability" components which came accompanied by test results and cost a fortune. They were the survivors.

Since waivers, deviations, and low yields were common, there was little probability that things were going to work. The longer it went on, the more "facts" the professional literature found to prove that it had to be this way.

However, this is a classic case of not being able to see the forest because the trees get in the way—the trees in this case being preconceived notions based on applying assumptions to scientific principles. We live with it all the time. For instance, Caesar was born surgically and so we call that method of childbirth the Caesarian section. However, his mother outlived him. Now do you seriously believe that in a time before Christ, the doctors could cut a woman open, take out the baby, sew her up, and she would live?

To change people's thinking on such basic concepts requires a lot of work. It is happening, but many are struggling.

The way I look at it is this: I would not want to work in an organization whose objective was less than zero defects.

Question 29 _____

What is the relationship between—or the difference between—the terms zero defects and no mistakes?

Answer

That is an interesting thought. They mean the same thing, but "no mistakes" sounds a little more negative. There are lots of ways of saying it: defect-free; without error; no mistakes; do it right the first time (DIRTFT)—dirtfoot, if you prefer; no failure. What is important is to use something that people just cannot misunderstand, a term that there is no way to get around.

Question 30 _____

What is zero defects in golf?

Answer

The same as it is in everything else: following the rules—playing according to the tenets of the game. Score has nothing to do with it, as is true in other sports where the individual performs.

In golf, as in no other sport, the participants on the whole do not follow the rules and in many cases do not even know about them. So except for the professionals and low-handicap amateurs, everyone plays to the rules that are convenient for themselves. They do not count every shot, nor do they take every penalty.

They throw down another ball rather than going back to hit again after losing their first ball. They give each other putts—"gimme's." It is all agreed upon and it is fun, but it is not golf.

Question 31 _____

How does the concept of zero defects fit in with the 14-step implementation process you describe?

Answer

Zero defects is the objective. The 14 steps are the actions that need to be taken to move an outfit along. I like to call it "the yellow brick road" [see Question 53].

There is no use tromping along some road unless there is an objective in mind. To deal with "lions, tigers, and bears," to fight off evil witches, to face the terrible wizard—all of these actions have no meaning without a worthwhile objective. We have to want to achieve something that improves the satisfaction we have with life.

The reality of the work world is that very few people actually enjoy work. There is a great deal of frustration and an enormous lack of satisfaction for the great masses of working people. Much of this is due to their determination that the specific tasks they do are not important or meaningful. There is so much checking going on; there is so little that can be depended upon; management is so insensitive to what people feel or want.

A company's purpose is to give people worthwhile lives by providing the opportunity for meaningful work, a decent living, and an opportunity to make a contribution to others. In pursuit of this purpose, a company has to make money, it has to grow, and it has to be successful. That can only happen when employees, suppliers, and customers all are singing the same tune. And that tune just can't be: "Two out of three ain't bad."

Question 32 _____

We've had a lot of discussion on the implementation of the 14 steps in quality improvement, the timing of the implementation of the various steps, and, in particular, "zero-defects day" and "error-cause removal." What is your view regarding the timing for implementing the steps?

Answer

The timing of the steps varies with every situation. Error-cause removal is something that can be done whenever the system is ready to receive and act upon the responses that will be generated. Zero-defects day is something that is done when management is ready to stand up in front of everyone and make their commitment clear.

The 14 steps of quality improvement were never intended to be a calendar. Back when I was still at Martin, I found that people accepted my philosophy of prevention with little difficulty. But the next question always related to "how" and "what" to do. So I sat down and wrote out a list of things that required something be done about them. When I was done, there were 13 items so I added "do it over again" in order to avoid having what some would consider an unlucky number.

The steps are taught in sequence only because they need to be discussed in some logical fashion. However, Step 8, Education, actually begins even before Management Commitment, which is Step 1.

The whole improvement team process is a learning event. If everyone in the company could serve on a team at some time, there never would be any problem of understanding.

When I went to ITT, I had all of this figured out and had even written a book on it. I prepared a brief brochure entitled

"Quality Improvement through Defect Prevention." As part of this I had a cassette made which recited the concept of zero defects in seven languages.

The brochure and the cassette were placed in a handkerchief-type box and mailed out, with a personal letter from me to plant managers and up through the world of ITT. To this very day I have yet to hear from anyone about those boxes. So I decided that each of those people had to be converted personally.

As we learned to teach the 14 steps, we struggled with the question of how to let everyone know firmly that they had to do all of them, but at the same time not imply some ritualistic activity that would stifle it all. If we don't insist, for instance, many will not have a zero-defects day.

Some feel that such an event, with the gathering of people, excitement, and such would embarrass them. However, I have been to dozens of them and everyone is always glad that they took place. I get invited to many more than I could possibly attend so we made a brief film, which qualifying companies receive, in which I talk to the group from an informal setting. If I had my way, I would go to them all. They are wonderful.

Similarly, in *Quality Without Tears* I wrote about the "vaccine ingredients," which are really advanced actions necessary to lock in quality improvement. In order to avoid discussions on the subject, I was very careful not to number them. However, as soon as the book came out, people began to count.

Question 33 _____

A few years ago, you were quoted in a newspaper as saying that shoddy workmanship can cost a company 25 percent of its annual sales, and if it is a service company, as much as 35 percent of its annual operating budget. Where do these num-

*bers come from? Are they still true? Were you referring to big
companies, little companies, or across the board?*

Answer

Those numbers are low, particularly in the service areas. I figure that every other person in service companies spends 100
percent of his or her time doing things over, chasing after
data, or apologizing to someone. The "manufacturing" companies really are mostly service companies too. They have a
small percentage of their people who actually put the product
together. The rest of the work force is doing something with
paper.

Each company is a process from one end to the other. All
work is a process. Everything is linked together in a series circuit. Whenever anything goes wrong in one area, the shock
waves are felt throughout the organization. That is why it is so
important for everyone to be involved in efforts to do things
right.

A great analogy is the health care center, like a hospital.
There, you have professionals such as surgeons doing one
thing and professionals such as accountants doing something
else. More people get mad at the accountants than get mad at
the surgeons. But the facts of life are that there is a consistent
thread of error permeating hospitals that don't take measures
to correct problems.

Studies show that laboratory tests, which have a built-in
statistical error, are unreliable by as much as 50 percent because of operator error. Where the error is discovered, the
tests have to be rerun. Where it is not discovered, there may
or may not be a lot of trouble as a result. Procedural errors
happen regularly in other areas as well.

You would be hard pressed to find a more dedicated group
of people than those who work in health care. I know a lot of
them and without exception they want everything to be cor-

rect. It is just that they don't instinctively understand how to make that happen.

Once a price of nonconformance has been determined for an organization, and even a phase-zero price of nonconformance is startling, then they realize they have to do something. That is the whole reason for figuring out such a price in the first place. I don't even tell people what I think the real numbers are because they would think I was exaggerating.

Question 34 _____

Why did you choose to say the "price" of nonconformance instead of the "cost" of nonconformance?

Answer

Price seemed to me to be a better word because of the way it is used. The dictionary meaning of price is cost, but people look at price as what you pay, whereas cost is viewed as what someone sets it to be. So to me price gave the opportunity to teach that people had a choice: "The cost may be X, but I am only willing to pay this price."

I was also trying to get away from all of the baggage that went along with the phrase "cost of quality." This has been so misused and misapplied that I wanted to get rid of it. Instead of talking about what a company was spending to do things wrong, people would argue about how this or that item was to be classified, whether under the category of failure, appraisal, or prevention. Nothing was getting done.

Question 35 _____

Rather than using a word like quality to describe the end that is sought, but which means different things to different people,

why not use something more specific, such as nonconformance,
which connotes measurability?

Answer

That might be a good word. Integrity has gained some use
lately, also. The problem lies deep in the perception that people have about goodness. It is very hard to get people to use
old words differently. I have taken the approach of trying to
get them to understand that if quality is to be managed, it
must have a meaning that is manageable. One more new
word probably wouldn't be of much help. It probably would
add to the confusion.

"Excellence" has confused people, for example. Everyone
uses it as if they had a common understanding of what it
means. Teary-eyed speakers yearn for its accomplishment.
Corporations etch it in bronze across their doorways. Yet when
pinned down, these same people cannot explain what one has to
do in order to be excellent. It is because they haven't really
thought about it. A list can be made, specifics can be defined,
and the requirements for being excellent, in the eyes of those
beholders, can, as a result, be laid out for all to see.

But people don't like to do all that work. They want one word
that encompasses all aspects of a field. "Productivity" is a word
that sounds like everyone knows what it means. But try to get
agreement on it and you'll see the truth about such terms.

"Quality" as used in quality control and quality assurance
has always meant goodness because people were permitted to
make value judgments every day. "Goodness" was really what
it was all about. However, when the thought of conformance
and nonconformance begins to permeate the operation,
"goodness" suddenly becomes inadequate. The search goes
out for something else.

I have tried to place the emphasis on requirements rather
than anything else. This point is easily missed. When some

talk about "fitness for use," they want to use that phrase to permit them to make judgments on products or services. They want to use it as permission to make waivers or deviations. Yet that was not its intent. It merely means that the output should do the job for the customer and should have requirements that spell it out clearly so it is "fit for use."

All requirements come from the customer in one form or another, because with no customers there is no business. But there are many customers besides the actual user. There is the Internal Revenue Service, which has its requirements; the SEC; the bank; the union; the landlord; and so on. All of these want conformance to requirements from us. They do not appreciate deviations.

There is a story about President Truman making a speech to a group of fertilizer manufacturers. He kept referring to "dung" as he talked about their product. After it was over, someone asked Mrs. Truman if she could encourage the President to say fertilizer instead of dung.

"It's taken me 25 years to get him to say dung," she replied.

I have been working longer than that to get people to understand the word "quality" properly.

Question 36 _____

The quality improvement literature we've been given says that defect-free performance is not the same as perfectionism. I'm concerned about employees getting that as a message for every project they're assigned. Could you expound a little on the difference?

Answer

The Red Queen in *Alice in Wonderland* was able to deal with communication problems quite easily. In her realm, words meant exactly what she wanted them to mean. The rest of us, though, have to explain our meanings to everyone. That is the purpose of quality education.

The dictionary meaning of "perfect" is "without defect or fault." However, common usage is much stronger than that, like having the power of healing along with no faults.

Any word is okay if it is explained. The word "excellence" gets a lot of use, but until we lay out the criteria for excellence, no one really understands what is meant. "Customer requirements" has meaning when they are listed one at a time and then the steps necessary to meet them are developed. Meeting the requirements then produces the result we are after.

The larger issue here is recognizing that all words have specific meanings. When we use them to communicate, we have to take time to make certain we are all singing the same song. The situation comedies on TV build many of their shows on the misunderstandings that happen between people who live in the same house, speak the same language, and have the same goals.

Understanding things in a similar way is a key part of the quality improvement process. That is why all the people in any given company need to have an education that explains their personal role to them.

The proper usage of words in communication is very important in an organization. When imprecise words are used, there is confusion and dilution of effort. The proper requirements make everything possible.

For instance, I'll bet a lot of people here were Boy Scouts or Girl Scouts. I'll also bet that all who were remember the Scout Law and the Oath. Now those are requirements that a young person can understand. In the handbook there are ten laws with a discussion about each one and a drawing of a Scout obeying each law.

Cleanliness is made clear by seeing Scouts bathing regularly. Helpfulness is illustrated by showing a Scout leading an elderly person across the street. Trustworthiness, reverence, obedience, and all the rest are explained and discussed thoroughly. On top of these conceptual basics, the Scouts have task-oriented activities through which they can earn merit badges. Thus concepts and implementation come together.

These young people learn how to get along with each other, how to set and achieve goals, and as a result, they have a good entry into adult life. When companies are willing to work that hard at causing understanding, then they will be able to cause quality to happen on a regular basis.

Question 37

Since the original conception of your philosophy, you seem to have evolved more and more systems. Are new systems replacing other older ones? Do you worry about ending up with a plethora of systems for achieving your end?

Answer

We only have the one process, and it is a complete system. I certainly don't plan on a bunch of different systems. You are probably referring to such things as the "quality vaccine serum ingredients." I wrote these out, not as a system, but in the hope that they would show the specifics that have to happen in order to cause quality improvement all the time.

I keep trying to think up additional ways of explaining all the concepts involved in quality improvement because I want people to really understand the process clearly. So when people ask me for specific things, I try to give them. Now the "vaccine" was created in response to someone asking how to get past the enlightenment stage on the maturity grid [see *Quality Is Free*]. They wanted to know who had to do what so I wrote some things down. These ingredients are divided into five areas: integrity, systems (meaning functional department

activities that support the quality improvement process), communications, operations, and policies.

They are like the operating instructions that come with a
new car. They don't teach the craft of driving; they tell about
the new machine. They let you know what is important. In
the same sense we have added many tools to support the process. Statistical process control courses and implementation
methods are one area; prevention implementation for managers—to teach them how to cause prevention; and prevention
implementation for individuals, which is the quality education system graduate course. But these are not systems, they
are educational programs that support the basic concepts and
amplify what has already been learned by supplying more sophisticated tools. More "systems" are not needed. Harder work
and a more serious application are what require attention.

However, we are often asked to do things outside the quality field, and we respond to the clients in such matters. If
quality is understood, the other aspects work themselves out.
Figuring out what to do in order to run a business is not the
difficult part. Getting people to do what you have figured out
is where the wicket gets stickey.

Question 38 _____

*How does your philosophy of quality differ from those of Dr.
Deming and the other "quality gurus?"*

Answer

I find this "guru" business tiresome and I think the others do
too. We have all spent our lifetimes trying to lead business in
from the wilderness and now find ourselves sitting on a cold
rock in front of a cave.

Last month I was walking through an airport and someone
yelled, "Say something in guru." So I raised my hand, palm

out, to him and said, "Do not rub Ben Gay on everything, my son."

It is not possible to take people with as much experience as Dr. Deming, Dr. Juran, and I and put us in boxes with clear labels like in a zoo. We all believe that the problem of quality belongs to management. We all believe that prevention is the way to get it. And we are all impatient that everyone is not leaping into what we see as a sensible, mature philosophy of doing things.

Dr. Deming has emphasized statistics over the years and has taught that approach to thousands of people. Dr. Juran is known for his quality engineering methods. If you do what they teach, you will do very well. They are dedicated people and worthy of respect. Dr. Deming and I write notes back and forth. Dr. Juran seems to think I am a charlatan and hasn't missed many opportunities to say that over the years.

The differences we have philosophically are not completely insignificant, though. I believe zero defects is a practical, reasonable, and readily achievable goal. They both seem to be coming around to that. In implementation, I have always aimed what I do at management rather than at the quality control people. This is primarily because management doesn't know what to do about the problem. So we have set up what we call the "Crosby Complete Quality Management System," which supplies educational material for every level in the quality improvement process, supplies the statistical tools, the corrective action, the supplier management, and, in fact, everything that will ever be needed.

Developing this implementation material required a lot of money and the efforts of a large professional staff. The teaching activities of this staff are supplemented with full-length case history films and numerous other teaching tools. We spend a fortune training our people. The idea has been to build a solid firm that will last long after I and the other individuals involved now are gone.

Another big difference is that I was a practitioner in the field for 25 years and a senior executive of a large multinational, diversified corporation for 14 of those years. I have not been a college professor. I have been there with hands on at every level of the organization from inspector to CEO. Because of this, I feel I can relate to the specific problems and objectives of those who have to get the job done. For instance, we teach statistical process control and supply the software to implement it. But in doing so we begin by explaining to management how it should be understood and used for continuous improvement. They don't need to learn how to make the charts or figure the numbers. They need to know what to do with it. Then the people who are applying what we taught them in the operation can communicate with the management. Over the years management has tended only to learn how to say "hmmmm" when presented with a chart.

The quality mission involves helping companies develop an overall strategy that will serve them for years, and this must be continually updated. I can illustrate by explaining how this is done at PCA. We try to put companies in the quality management business. Several companies, such as Johnson & Johnson, General Motors, Chrysler, and Federal Prison Industries have set up their own quality colleges under our supervision. We take their instructors into our operation for six months or so and they learn how to teach the management college and later we teach them other courses.

So we are involved in helping companies change their culture for the better permanently. Lasting quality improvement is much more than just providing the tools of quality control.

I do not consider that we are in the same business as others in the quality field. We have spent untold dollars on materials and much more on facilities. The Quality College is an institution, not a lecture series.

Question 39 _____

In the past 12 years my company has gone through product line planning; we've gone through quality of work life and quality circles; we've gone through MRP. Now it's just-in-time and SEM. Every time the "evangelist" leaves, there is difficulty in ongoing implementation of the process. What would you suggest?

Answer

Your question is good support for my "elf" theory. Management installs programs to solve problems on an ad hoc basis or to move the company along instead of adjusting the company culture to meet the realities of the marketplace [see Question 40].

No system, including the one I and my associates have devised, works all by itself. Management has to understand what the system is and has to take part in making it a living, breathing part of the operation.

That is why it is fruitless to try to talk people into installing a quality improvement process. They have to want it and be willing to work for it. Once begun, the quality improvement process helps those other programs you mentioned because it serves as an umbrella under which they can blossom.

It is all sort of like the exercise part of a wellness program. No one can come in and move my muscles for me, or force my lungs to expand and contract, or lug me on a stimulating walk. I can pay others to do the exercises instead of me and mark it down on the card. Then I will receive my credits, but I will atrophy while it is going on.

Many of us have financial advisers to help us deal with the complex world of tax and investment. However, we are the ones who must put the advice to work. The world has many people who turned it all over to someone else and paid the price.

The success of the quality improvement process does not depend on any "evangelical" powers possessed by the quality experts. It depends on education and implementation conducted in a serious and methodical way. Many people think that if they pay attention to their spiritual life by attending worship for one hour a week, everything will be okay. However, our personal spiritual life is not something to be handed off to a church or any religious organization. It is up to us to attend to it. The organization provides fellowship, teaching, and counseling. That is support. We have to bell the cat ourselves; others can't get us into heaven.

So what is going to happen in your company with the installation of the quality improvement process depends very much on the attitude of management and the participation by the employees. Experience shows clearly that companies can turn around and prosper with little expense and not that great an effort. But the culture has to change. Programs will not accomplish that.

The quality improvement process serves as an umbrella that lets the organization conduct programs designed to assist specific functions. Activities like just-in-time inventory, work groups, management by objective, employee education, and many others will have a better reception and implementation because of the quality improvement process.

Best of all, a quality improvement process works forever when it is installed properly.

Question 40 _____

How do you view the management by objective program in the quality improvement process?

Answer

Management by objective, often called MBO, is a concept originally put forth by Dr. Peter Drucker, who, incidentally, is a personal hero of mine. My thought is that he saw it as a perfectly natural system of attacking a problem. It has been laid out as some sort of master scheme but is still a useful way of doing things. It fits right in with the quality improvement process that I advocate.

Suppose, for example, the decision is made to eliminate customer complaints within 12 months. There are 60 complaints a month now, so the plan is to drop by five a month in order to meet the objective. To do that, it is necessary to have many subordinate objectives that involve identifying the causes of the complaints, defining their characteristics, and eliminating their source. Everyone has a job to do, with clear methods of measurement and a completion schedule.

All of this can be laid out in a master page on line sheets and tracked through meetings. It is a formal system that works very well. The key difference that the quality improvement process brings is that instead of reducing problems, the goal becomes to eliminate them forever.

One nice thing about the entire quality improvement process is that it acts as an umbrella for the whole company. You can hang almost every kind of program under it effectively.

Of course, you have to be careful because management just loves to find packages that replace thought and original work. There is an incredible amount of mythology in management thinking. Much of it originates in the business schools. I talk to them regularly and find the students open-minded while the professors, with some exceptions, still think quality is a technical part of the system and that it is a problem of the workers.

Recently I took my seven-year-old grandson to see David Copperfield, the magician. It was a great show of illusion and Copperfield is a master at it. At the end of the show, he called a man down from the audience and gave him an old oak bucket. Then while the man held the tub, Copperfield carried a duck across the stage and put it in an enclosed box. When he opened the box and took it apart completely, there was no duck. He walked over to the the old oak bucket, reached in, and pulled out the duck. My grandson Charlie cheered and so did everyone else.

Then a voice—planted, I am certain, by the star—called out: "Do it over again, in slow motion."

Copperfield called the tub holder back and showed everyone that it was still empty. Then he went to the side of the stage and came back with a stuffed duck which he took over to the reassembled box. As he started to put the duck in the box a man appeared stage-left dressed as Mercury. He had a tin hat with wings, wings on his heels, and was wearing a skintight leotard. Mercury came in slow motion across the stage and took the duck from Copperfield. Then both of them went over to the tub and stuffed, with some difficulty, the false duck down into the barrel.

After Mercury went off stage to thunderous applause, the magician reached into the bucket and pulled out a live duck. The audience thought this was wonderful, as did Charlie.

On the way across the parking lot on the way out, I asked Charlie if he thought he could do that thing with the duck. He looked up at me with a patient expression and stated: "They showed you how to do it; all you need is one of those elfs."

Well, that is the way most executives manage. If the marketing operation is not doing so well, get a new marketing elf. Quality a problem? Try this new elf dust called quality circles, go to this seminar, get a new quality manager.

It is no wonder people get confused. That is why I have spent my career trying to make things simple enough so that I could understand them, and then pass the information along.

Question 41 _____

My company is in the midst of beginning a just-in-time implementation and we are looking at the zero-defects program to complement that. Do you see any potential conflict of interest between those two efforts? Do you have any watchwords for us?

Answer

The essential ingredient of a "just-in-time" inventory program is receiving defect-free material and services from suppliers. It is easy to have things delivered frequently, but if they are not correct, then delays will certainly screw up the whole operation. The suppliers will wind up having inventory near the plant to make up for that. Then the "less expensive" goal does not happen.

Not only that but if the system is not well thought out and the suppliers are not properly oriented, it creates a very difficult situation. Some large plants have traffic controllers working on the CBs to keep all their trucks straight. Some suppliers make a delivery, then go park under a tree across the street to wait for 3:00 p.m. Just-in-time can be very difficult if the thinking is short range.

The key to making it all happen is getting the suppliers interested in understanding and then meeting the requirements. After that comes establishing a deviation-free environment inside your operation. Don't depend on some elf to handle it for you; a system is not enough to make it happen.

Just-in-time really started with Henry Ford in the early 1920s. He set up suppliers around the Rouge plant and they would deliver as needed. The Japanese picked it up for a different reason. They didn't have material that could be wasted like we did. So they figured that if you want a person to do ten, you give them ten sets of components. We are used to providing a 10 to 20 percent excess in the name of efficiency. Just-in-time is a complex issue, not something that is to be taken lightly.

Question 42 _____

In one section of Running Things *you expressed some different ideas on employee performance appraisals. Would you expand on your view of such appraisals?*

Answer

Running Things had a chapter on the trade-offs of people who
are effective on the one hand and drive you nuts on the other.
When the lines cross, the "hassle factor" comes into play and
the person has to get straight or get out. As a result of that, I
received so much "you're-so-smart; you-tell-us-a-better-
system-of-performance-appraisal" hassle that I put a complete
system in the next book, *The Eternally Successful Organization*.

The basic idea is that the employee be given real input to
the system, including evaluating the management. It always
bugged me that someone I didn't choose got to make decisions about my career. Only one person I ever worked for
in my early years thought I was capable of more responsibility. I usually received the "we-don't-think-you-really-are-the-
person-for-the-job-but-we-don't-have-anyone-else" routine.

In *The Eternally Successful Organization* I wrote about a
fictional company, the Masters Corporation, which was trying
to create a performance review system that would actually
contribute to the individual as well as to the company. The
solution involved more than making out a new form and providing something for the human resources department to
evaluate. It is a whole new way of looking at performance review, and it is, in real life, under development for companies
interested in taking it on.

Performance appraisal is one of the most important things
managers can do, but hardly anyone does it well. The system
gets tied up in merit increases rather than personal development. When a development effort is made, it is usually aimed
at the wrong people because there is no really accurate measurement system for selecting prospects for development.

The only good performance review I ever had was put together by Jim Halpin, my boss at Martin. Jim was a wonderful
executive. He picked me from the bowels of the organization
and gave me my first big shot at anything. What he did was

pull together the six or so managers whom I dealt with regularly: my other boss, the project director, and the people who ran manufacturing, engineering, purchasing, marketing, and a couple of other "ings." They had a no-holds-barred discussion about me, which Halpin summarized in a report. We then discussed the report.

What I learned was how I appeared to people, some changes I needed to make in my relating and managing style, and some places where my communication skills were inadequate. It was a great thing and I carried a copy around with me for years. As far as I know, no one ever did that for anyone else at Martin. I tried it with a couple of my people, but they did not appreciate others making comments about them. It is a more difficult subject to pull off smoothly than one might think.

Question 43 _____

In your book Quality Is Free *and in your work of spreading quality concepts around the world, you and your organization have been very successful. What future activities do you have planned to continue airing your message?*

Answer

I like your question. We are still in the process of learning how to reach the hearts and minds of people in business. In eight years PCA has grown into a company of 250 people, with operations in London, Paris, Munich, Florence, Sydney, Singapore, Tokyo, San Jose, and Deerfield, Illinois, in addition to Winter Park.

To reach people requires spending a lot of time and money training people and outfitting factilities. Overseas efforts, of course, mean translating material into several languages and learning how to manage a business on an international level.

The professional part, the message itself, has been easier to create and disseminate than has been the actual management of day-to-day affairs. International financial operations are incredibly complex and are easy to underestimate.

The interesting thing about all this is that I am really more a writer than a businessman. I started this company and helped run it in order to provide a platform and a vehicle for helping get the world straight on quality. I tried going from lecture hall to lecture hall, but that doesn't reach people very effectively. They need an immersion.

My personal plan for the future is to continue refining a quality philosophy and trying to communicate it, to make speeches to client groups, to work with clients, and to help PCA's management with strategy. My wife and I are building a new home in Winter Park, part of which is a library. That is where you will find me most of the time.

The Crosby organization will continue to grow and to develop ways of assisting companies with positive change. Most of our emphasis is on helping companies set up quality management systems on a permanent basis, without the need for day-to-day support from an outside source after the initial implementation. That way they can benefit forever. We want any organization needing help in instituting a quality improvement process to be able to get it without a lot of hassle. Our corporate policy goal is to make prevention of defects a standard management approach throughout the world.

Question 44 _____

Are the case histories you use in your books real? Do you do the writing yourself?

Answer

I do the writing. I consider myself a writer first. The rest of this stuff is just to earn a living.

The case histories are "living fiction." They are based on actual life experiences, but they represent a composite of many situations and events. That way, I can arrange events to come out the way that best makes the point, although I have to admit that most of the time I don't know how a case study is going to unfold once it's begun. For example, the long case of the Masters Corporation in *The Eternally Successful Organization* started out as only two chapters and wound up to be half the book. It took on a life of its own and came to include a corporate strategy meeting and original solutions to some very complex problems.

Cases are a good way of telling the story. In *The Art of Getting Your Own Sweet Way*, I used this technique a lot. One hero was a church program coordinator who solved such problems as where to get Sunday school teachers and how to deal with the church dinner when the women in the auxiliaries refused to cook. Then there was the little league president who didn't want to do everything himself; the up-and-coming young executive who had to convince an old-timer to accept a new idea; an association president trying to deal with a board that didn't want to do anything; a team trying to start a new plant in a new location; a staff member trying to write a bad news status report to a boss who didn't like bad news; and one on the "family executive," who usually turns out to be the woman of the house.

In *Quality Is Free* I use the HPA case to take a team through the whole 14-step process. There is also a visit to a decaying hotel and a quality engineer trying to transform the shop manufacturing manager from Mr. Hyde back into Dr. Jekyll. It had several integrated cases.

Quality Without Tears contains a takeoff on Dickens, "The Quality Carol," in which a thinly disguised Scrooge learns he is destined to spend eternity fixing all the defective things he has shipped out in order to meet schedule. We made a film of that one starring Ephraim Zimbalist, Jr. The Lightblue Corporation's planning system was examined in that case, and both of these had a chapter in the back of the book showing how it all came out. Other smaller cases were included.

Running Things has many cases also. "The GNU Story" showed a person starting his own company and examined the situations he had to handle in order to get it up and running. "The Three Little Pyggs" struggled with policy setting in their new ventures; an executive committee met to abuse those they represented; and the ants straightened out a closet grasshopper.

I find that explaining a concept through a case study makes it easier for the reader to digest. The kinds of subjects I write about can get ponderous in a hurry when the writer's word is all there is to choose from. The cases are an attempt to lighten the book up, make it more entertaining. I think there is no reason to ever be dull if you can figure out a way to avoid it. For these same reasons, I try to avoid long words and technical terminology in my material.

Question 45 _____

Speaking of learning, how do you personally keep up to date and keep learning? What are your primary sources?

Answer

I read a lot of biographies. Biographies and history books keep me informed on how people responded when faced with certain situations. The names of the situations change over the years but the content remains quite similar. I find such books to be a constant help in determining strategy and keeping things in perspective. Our company has been through a lot in the past year: an embezzlement; business problems for a couple of major clients; international accounting mixups; the falling dollar—which makes profitability difficult because the value of inventory overseas keeps going up even though the amount of material there does not change; negotiations for a merger which took several months and then came to nothing as both sides thought better of it; harassment from stock analysts who don't understand that it costs money to translate current material, develop new material, and train international consultants. Difficult situations are all just part of doing business and are overcome by not losing perspective and by diligently dealing with them.

To keep up with what is going on from day to day, I read *The New York Times, The Wall Street Journal, The Orlando Sentinel*, and if I'm on the road, *U.S.A. Today*. I listen to local news and watch CNN on the TV. I read magazines, among them *Time, Forbes, Fortune, Inc., Harpers, Atlantic, The New Yorker, New York, M, Town and Country, Central Florida Magazine, Orbus*, and a few others.

I have learned that it is necessary to be interested in everything because one never knows when something will strike a spark. Building up a large data bank in one's head provides a source of understanding about situations. Then the thought process will work its way around to something practical and usable. Most of my works have originated from a thought scribbled on a scrap of paper and shoved into a pocket. The receiver always has to be on.

People are always going off to college classes at night in order to learn something. I did that years ago, of course, and still go to hear others speak and comment. But the best educator of yourself is yourself as long as you don't get too terribly specific.

Question 46 _____

Would you recount the filming of The Quality Man?

Answer

The British Broadcasting Company filmed it at Muirfield, Scotland, which is the oldest golf course around. The British

Open was played there in 1987 and, of course, many times before.

The idea for the film came about when a BBC executive read about us in a *Time* article and dispatched a producer to check us out. A gentleman named Brian Davies came over and spent three weeks living with the company and clients. Then he went home and wrote me concerning what he would like to do, which was to shoot about two days of me talking and then edit it down to 30 minutes or so. They would sell the film for a while in hopes of getting their money back and then play it on BBC. He asked me to come to the U.K. for it and said they would pay me what amounted to $300.

I figured it would cost around $4000 for me to make the trip and spend the time, but the executive committee thought it could be worked out well if I would do it during the vacation I had planned for Scotland that year, 1985. They volunteered to pay for my room and board during the filming.

I told Brian that I was not enthusiastic about giving up some vacation time but would do it if he could get me on Muirfield, to which even the big pros have a hard time gaining entrance. The Honorable Edinburgh Golfers do not suffer tourists well.

He got it arranged and had to pay a fee. So we all met there, he with a crew and Peggy and I with clubs. We filmed for two days of the most excellent weather they had ever had, and I got to play a total of nine holes. It is not easy with six people around you saying "hold it," "just a moment," "okay, now hit it." However, it was a fun experience.

The BBC in Europe and Films, Inc. from Chicago started to market the film and to their surprise sold a lot of them. In fact, sales have been so good for them that they never got around to running it on television until mid-1988. They were afraid that people would copy it on VCRs and that would put them out of business. I suspect it will be seen on Public Broadcasting soon.

It shows what a producer/director can do when he really

understands the subject and the "actor." I just did what he said and it all worked out very well. They made a lot of money, I lost three days of my vacation, PCA got $300, but it has been good for PR.

Question 47 _____

Where do you get the case history films that you use in the Quality College classes?

Answer

We have a creative department that does these things. For the first few years I was the creative department, but now we are better organized. We do not have our own film company, but we deal with two or three that know what we want, and we use professional actors and crews to make certain that what is provided is something that will hold the students' attention.

The case histories are worked out to follow real-life and actual experiences and at the same time to deliver the message that knits the entire week's work into a coherent pattern. The objective is to take a class consisting of people from different companies or work areas and give them a common reference point.

The idea for a film is, of course, the starting point. It is then necessary to meet with the producer, writer, and project manager to kick it around. We try to find something interesting to do that will keep the participants involved. Cases like the Magnabank case take a lot of research in order to be judged authentic. We spend many hours talking to people in an attempt to be as accurate as possible.

As the writer develops the script, it is essential to make certain the concepts are being treated properly and that it fits the students. When the script is complete, I stay out of production except to perform any role I might have in the film. We try to make them wear well. Films are very expensive, but they are so worthwhile that we consider it worth the price. One thing we hadn't counted on was translation for overseas use. It is much harder than one would think and costs about $500,000 per language for the films and the other material. It also takes a lot of time and effort. But the kinds of case history films used in our quality education process communicate our philosophy much more effectively than would be possible just using talking heads.

Question 48

Why do you use so many grids?

Answer

There are really only three, but they are good conversational points. The Maturity Grid from *Quality Is Free* was set up originally to make the point that the quality improvement process is progressive. One doesn't just go from awful to wonderful in one leap.

In *Quality Without Tears* I used a profile, which is a kind of grid, for the same purpose. It is much more mature than the first one and gives people a quick check on where the company is at any given moment in accomplishing a specific part of the process.

The third grid is in *The Eternally Successful Organization*. It covers a company's overall progression rather than developments in the quality area exclusively. The grid elements relate to various stages of health: comatose, intensive care, progressive care, healing, and wellness.

All of these grids make it possible to refer to grid elements or stages throughout a book without having to go back and explain each one every time. They make a great reference point and provide a way of measuring where different people think operations stand. They provide the opportunity for a performance rating that can be done by anyone who has some experience with the operation.

If you have the chance, ask several different people, including a customer or two, to rate the company's operation on any of the the the grids mentioned. This exercise will provide enough data to encourage even more attention to improvement.

Question 49 _____

How do you go about selecting the titles for the books you write?

Answer

All of them come by chance. In 1972 I wrote a book called *The Strategy of Situation Management*, which the publisher wanted to publish but they didn't like the title. I loved it. However, I agreed to keep thinking. Then one night I went to see a revival on Broadway of a musical called *No, No, Nannette* (starring Ruby Keeler). At one point in the play, one of the actors said, "You can't always have your own sweet way." I remembered my grandmother saying that to me years ago and I stood up right there and said, "That's it." So we now have the first and second editions of *The Art of Getting Your Own Sweet Way*.

With *Quality Is Free* I had a hard time keeping the words "quality control" out of the title. The publisher wanted them in there. I feared they would kill the book's appeal to a management audience. Then one day the CEO of the company where I was working said, "Quality doesn't cost anything, it is free." So I stole that line from him, with his permission. I made the subtitle *The Art of Making Quality Certain*, which is what I had wanted the main title to be all along.

I really couldn't think of a name for the next book. I wanted to call it *The Art of Hassle-Free Management*, but the publisher thought no one would know what the book was about. My editor and I had a lengthy discussion and in kidding around he said that if I could incorporate sex and dieting into the same book, it would certainly be a hit. "We could call it *Sex and Dieting Without Tears*," I said. From there it was just one more step to *Quality Without Tears*.

Running Things I thought up without any outside inspiration. I decided to not use "the art of . . ." in any more titles. But when the book came out, there was the subtitle: *The Art of Making Things Happen*. It is a good book on management, not so much about quality, and it has a lot of case histories.

The title for *The Eternally Successful Organization* was a natural tie-in with the themes of wellness and prevention that are so critical to business these days. Again, it has little to do with quality per se. The subtitle is *The Art of Corporate Wellness*.

Question 50 _____

It sounds like what you're trying to do is to establish a set of values and a process for accomplishing the quality mission and to get people to realize that the values and the process are, in a way, as important as accomplishing the mission. To what extent do you find the senior people you talk to have a different vision and values? What portion really recognize what you're trying to do?

Answer

The values we espouse are central to the mission. The process is the means for accomplishing it. They are very important. The people we deal with are thoughtful people. All of you are here because your company sent you, and they sent you because they feel that what we help you do fits in with the goals and "values" of the organization.

However, there are a great many management teams out there that are very self-centered and have lost touch with reality on the subject. They have an artifical set of values or a clouded vision that is taking their company away from success.

The ratio of those who feel I am an unrealistic dreamer and those who think I am a pragmatist is changing. It would be difficult to give exact figures. I get a lot less hassle now, and most of what I do get is on what has to be done to change. There is very little disagreement on the need to change.

One of the reasons I know this is that I have taken to having "alter calls" after my speeches. Perhaps an anecdote will shed some light on the issue. It is a practice in many southern churches and among most of the charismatic congregations to provide a time after their worship service for those who want to change their ways, or join the church, or ask for prayer, to come forward. It is considered a rite of passage in spiritual life that you have to stand up and be counted if you want to be in compliance with God's will.

Well, the quality mission is hardly the same thing, but there are parallels. I have been trying to reach audiences for so long, to get them to feel in their hearts that they have to change their wicked ways. So a few years ago when I was about finished speaking to a very large group in San Jose, I asked them to raise their hands if they were willing to go back and start a policy of delivering defect-free products and services to their clients, on time.

About 15 of the 700 raised their hands. So I talked about how this didn't have to be a policy for the whole corporation, just for them, and about how each of them could be a witness for quality improvement. This time about 80 of the 700 raised their hands. I was getting upset, so I started ranting and raving about how they were selling their companies and their country down the river, and how it was dishonest to advertise defect-free stuff and then not plan to deliver it.

This time I asked those who were willing to take that stand to rise to their feet. Well, about half of them did, and as they stood there the rest of them slowly joined, and before long the entire audience was up. I applauded them, and then they applauded themselves, and the speech was over.

But it stuck with me. There are a great many people who spend their time making up elaborate excuses for not doing what they have agreed to do. It is as dishonest as insider trading.

Quality Action: Toward a Workable Quality Process

Question 51

You give out "fanatics" cards in your classes, there is a "quality fanatic's" flag on the wall, and I've seen photographs of you presenting this flag to people. What is the story behind the fanatics?

Answer

It all goes back two years or so when the president of one of the companies which is implementing the quality improvement process said they were doing everything by the book but not enough was happening. I replied that they were just going through the motions and not putting enough energy into it. The process is a logical, tool-fitted way of managing but it doesn't do the job all by itself. "You have to be a fanatic about it," I said.

He asked me to be specific about what a quality fanatic did. The result is the card you refer to. When I see a company management that is performing according to the description on that card, I give them a plaque and a flag. We just began doing that recently so there are only a half dozen companies represented. Several more have been targeted for recognition as "fanatics."

But to be a fanatic, it is necessary to do a few things. This comes from the card.

First, decide you want a zero-defects strategy. (This means that the management has fully accepted the notion that doing things defect-free can become a way of life and they are going to do everything necessary to make it happen).

Announce a clear, specific quality policy. ("We will deliver defect-free products and services to our customers, both internal and external, on time." There should be no vacillation, no discussion of the "economics" of quality. It must be something that cannot be misunderstood. This policy has to be explained to everyone. It is essential that they understand what it means. I like to see it written out on the CEO's personal stationery, signed, and mounted in small frames right above the light switch next to every door in the organization).

Next, display management commitment through action. (Witnessing is not just standing up in front of everyone and being evangelical. The populace wants to see if management "walks like it talks." The tests start early when it comes time to deliver something and that something is not quite right. Whether the nonconforming product is a document or a truck, the people will know it is wrong. If management caves in to the pressures of time and money, then quality immediately crashes to its place as number three. If management holds firm, then nothing will ever reach that point in that condition again. It is as simple as that).

Assure that everyone is educated so they can perform. (Each and every person in the organization must understand his or her personal role in making quality happen. Employees need to be immersed in the absolutes of quality management, and they need the tools that will help them put prevention and problem solving into their jobs on a daily basis. But this education must be consistent. Everyone has to be able to understand each other on quality and do it automatically. We have companies tell us that quality is the only thing everyone can communicate on worldwide. Don't leave education to chance).

Then eliminate opportunities to compromise conformance. (Identify the procedures that show how to make waivers or deviations and tear them up. Invite everyone out to the parking lot to watch this destruction. As long as people have the ability to get around requirements, the temptation becomes irresistible. When I was a line-quality person, it didn't take long for me to realize that manufacturing knew that if I signed something off, it could be sent out whether it was correct or not. So they concentrated on working on the quality control people instead of on the product.

They can get very dramatic about it. They'll bring a little old lady into the shop and say: "This little old lady lives on the dividends that the company pays. If you don't sign off on this stuff, it won't be shipped, the company won't make any money, and this little old lady will starve." I had to learn to say, "Goodbye, little old lady.")

Insist that every supplier do the same. (Every company needs regular seminars for suppliers. Suppliers have to become an integral part of the operation and have to understand that they need to be fanatics also. The only way they will know this is for the purchasing company to tell them face to face. Bringing the senior people, not the sales staff, in for a seminar will start this in motion. In our company, for example, there is a workshop to show how to do this and provide films and material for support. It has to be done the right way and then backed up when things start arriving. It has to be recognized that it takes two to make a relationship and both have to listen. The buying operation has the responsiblity of making certain that the requirements are clear, the supplying organization has the responsiblity of making certain that they can deliver exactly that, when needed. It can be the beginning of a new world. If everything is done properly, it is possible to look forward to the elimination of receiving acceptance operations).

Convince everyone that they are all dependent on each other. (The awareness of the need for quality cannot be assumed; everyone has to know how this affects them personally. Employees need to understand where their work goes. They need measurements in order to keep track of how they are doing.

One of the perils that comes from an organized attack on some problem is that people will begin to feel that the system will take care of it all. Traffic lights all over the city help the traffic flow smoothly, but they do not eliminate the necessity of looking right and left before crossing the street. We have to

be educated about that and then take responsibility for it ourselves. We also have to know that others expect us to obey the
lights. They need to count on us stopping on red, so they will
be safe while progressing on green).

Satisfy the customer, first, last, and always. (This is not a
cliche; it is the ultimate reality. No one ever gets so wonderful
at what they do that they don't need customers. To do this, it
is necessary to listen to the customers and the rest of the marketplace).

F   First, decide you want a zero-defects strategy.

A   Announce a clear, specific quality policy.

N   Next, display management commitment through action.

A   Assure that everyone is educated so they can perform.

T   Then eliminate opportunities to compromise conformance.

I   Insist that every supplier do the same.

C   Convince everyone that they are dependent on each other.

S   Satisfy the customer, first, last, and always.

Fanatics have more fun than those who just work at a regular pace. Fanatics leave footprints instead of just dust.

Question 52 _____

*There is a widespread impression that your methods deal pri-
marily with motivating workers as the way to get improvement, but in learning the quality process, the word "motivation" hasn't really come up. Is that impression wrong?*

Answer

It is wrong. A couple of well-known leaders in the quality
consulting business formed the idea years ago that I was talking about "exhorting" workers to do better. So over the years

they have made that part of every speech they delivered or article they wrote. I have never responded, even though it should be fairly obvious to anyone who has ever read anything I have written or has heard me speak that I hold management responsible for the whole mess.

I, personally, and the company of which I am a part are very careful to avoid speaking ill of anyone, although we don't view any of those as being real-life competitors. I think that trying to torpedo the competition is not part of the business world. There is plenty of room for everyone. At PCA, we have had fewer than 100,000 executives and managers in our classrooms. At least 5 million need to come.

I see fewer and fewer cases of people writing and saying wrong things about my ideas. I hope that means the message is getting out. Motivation is part of every human exchange and it has to be considered. The recognition part of our process takes that into consideration. And sometimes special things apply.

For instance, I like to recognize outstanding performance with little gold stars that can be worn on the lapel or blouse, as the case may be. It is an attractive metal piece with my initials on it. They receive one when they send me a cost elimination idea, something they are going to do themselves. I paid for the stars myself. Everyone likes to have them and I enjoy giving them. That is a form of motivation that executives can use with their people.

Those skeptics I mentioned, for example, saw the Defense Department trying to reach people with bands and zero-defects days. They were critical and didn't stop to think that it was an attempt to do more about quality and should be encouraged. They just turned up their noses, turned off their minds, and kept on talking about how nothing can ever be right the first time.

Every profession has doomsayers like this, and it is possible to turn into one by being certain that what you believe is the absolute last word. We have to keep on learning every day.

Question 53 _____

You have a 14-step process for implementing your philosophy of quality. Can you summarize in one sentence or so each of the 14 steps in the quality process?

Answer

If I start doing that, I may wind up with six-page books, which no one will want. However, it is a fair challenge.

Step 1:   Management commitment is the willingness to give away something you cherish, something very personal, in order to improve the quality of other people's lives.

Step 2:   The quality improvement team is the "health care" group that is charged with supervising and coordinating the surgery, recovery, and wellness process in an organization.

Step 3:   Quality measurement is determining if the various "life support systems and procedures" are operating to the required results.

Step 4:   The cost of quality evaluation reveals the expense and inconvenience of doing things wrong.

Step 5:   Quality awareness is communicating continually in order to let everyone know they are on the same track.

Step 6:   Corrective action is identifying, curing, and then preventing the diseases that impair the enjoyment of life, be it personal or business.

Step 7:   Zero-defects planning is arranging for the day—commencement day—when management will stand up in front of everyone and declare that they have been converted.

Step 8:   Employee education involves building a base for

comprehension and implementation through a common language and the applicaton of special skills.

Step 9: Zero-defects day is the day when everyone gets together and celebrates their commitment to quality.

Step 10: Goal setting is describing the specific achievements that each individual is going to accomplish.

Step 11: Error cause removal is a system of pinpointing and eliminating the obstacles to zero defects.

Step 12: Recognition is acknowledgment, saying "thank you" to those who earn and deserve it.

Step 13: Quality councils are meetings of those responsible for an organization's wellness.

Step 14: Doing it over: To quote Albert Schweitzer, "Example is not the main thing in influencing others, it is the only thing."

Quality Without Tears has brief but very concise descriptions of these actions. But remember, they are guidelines, not instructions. Each one has to have something done about it.

Question 54 _____

I work in a large department (about 1700 employees) within a large company (about 42,000 people). How would you envision the scenario if one department tried to implement a quality process without having it fully accepted or understood by the rest of the company?

Answer

This is an opportunity to be a hero. Someone has to begin every march. As you and your team deal with the other people in the company, they will begin to realize that your department is different. When they discover that it is simpler to get requirements straight up front, they will begin to ask how you learned all this. It is a wonderful opportunity to influence and lead the corporation.

Most real changes start somewhere in a company and spread around because they are worthy. They do not come out of the boardroom. Think about the office computer systems and the complete revolution that has happened in the area of administration and control. People started using computers because it made their lives easier and let them do things that could not be done before. Then they began to want to tie them all together and the computer companies were forced to learn how to do that.

The management agreed to pay for all this and after a while realized that it was the only way to do business. But I know of no company, including the one I am a part of, where the initiative began with senior management. There are a couple of senior people here whom we don't even tell about computers. We just wire around them.

As the people of your department begin dealing with other members of the corporation, you will begin to receive questions concerning your way of operating. Others will see the difference in your department, that things are much more efficient and positive. Many departments in such a situation begin to share material and other information around the company. As other parts of the organization become more involved, the process will develop a momentum of its own.

Question 55 _____

In cases where a company's quality improvement team is being trained in two groups, is it better to start quality improvement as soon as the first group is trained or to wait until the remaining team members become trained and can join the group?

Answer

People come onto teams, people leave teams, whole teams change, but the process goes on. Everything has to begin sometime if it is to continue. Right now, there are people in every stage of the quality education, from the first grade through graduate school and beyond, so to speak. If we waited for everyone to be at the same level, we would never get anywhere.

When Winston Churchill became prime minister, he called Lord Ismay to the basement of a nearby building. As the two stood alone in this dingy room, Churchill said that he wanted Ismay to build a communications center in that room from which the war could be run.

"I'll help you roll up this carpet," said Churchill. And the two of them got down on the floor and started the process at that moment. Churchill didn't wait until the cleanup people came along. He wanted to get started, so they did.

For example, we go through continual training and education here inside our own company. For instance, we have account executives who are counselors/instructors. These folks are already experienced and successful in business when they join up. They have to be taught how to instruct in the College and how to go to an operation and help out. That takes seven to nine months of intense activity. They call it "boot camp." It is very hard work, but when they are through and certified, they can do all the things that an individual and a company in training will require.

Many people are in different stages of the educational process. We have several on-site right now from Europe and more coming. Also, we have educational steps for other levels of professionals, as well as those in the administrative ranks.

A quality improvement team is not a SWAT squad that has

to go solve problems. Quality improvement training is education of the team members, coordination between operations, and leadership of the process. It is a living entity that survives its members. It is not a ritual, it is leadership.

Question 56 _____

In a large company where all the employees have to be trained in this concept, and where it might take a year or more before it's finished, might some of those trained first need a refresher course by the time implementation begins?

Answer

Education is a continuing thing. People will want to attend some classes or information seminars on quality improvement all their lives. As I have already suggested, one cannot wait until the entire army is in perfect condition before facing the enemy. That is what General McClellan was doing in the first year of the War for the Union. President Lincoln asked if he might borrow the army if McClellan wasn't going to use it.

In the course of a business career, it is necessary to continually learn new things and study in order to stay current. It's the same in the quality process. That is why new techniques, products, and materials are perpetually under development. They are for the business community. People never grow tired of learning about things that help them live more interesting, more productive lives.

But such learning experiences and opportunities have to be scheduled. The best framework I've seen is a routine "show and tell" of what is happening. No one can hear enough success stories.

Many companies have adopted this practice. For example, Milliken, the large textile company and one of the first companies to go through our quality improvement process, calls these show and tell presentations "bragging sessions" and runs them monthly. Participants come to a session of the executive committee meeting and present the story of what they have accomplished. In turn they receive recognition in the form of certificates that acknowledge their achievements. These are marvelous sessions. "Show and tell" can be done at any level in an organization. It is an important part of the educational process.

Question 57

Can you share with us what your experience has been when an organization implements the process in only certain parts of its operation?

Answer

It is hard to do something in a corporation without others being involved eventually. What I have seen in every case that has been tried is that the parts that are not participating soon get very involved.

In General Motors, for instance, the plants that supply components to the assembly operations did not get involved in quality improvement as fast as the assembly plants did. When the assembly plants started refusing to accept supplies that did not conform to their requirements, these supplier operations began to get very interested. And when headquarters supported those who had been sent the poor material, it hit home. Now they are not only participating, they are leading the charge.

Things happen when they are supposed to happen. The civil rights events of the mid-1960s had been simmering for years. But what got it going was not the politicians or the activist groups. No, what made it happen was one elderly woman who refused to take another seat on a bus. When she put her foot down, people started to move toward doing what they knew they should have been doing all along.

It is the same in any company. It only takes one person, one division, one group to change the whole company. Every operation should have the opportunity and encouragement to participate. If they do not do this right away, then they must be expected to show improvement anyway. It doesn't take very long for them to get the idea.

Question 58 _____

In addition to what has already been given, what advice could you offer to someone who's having difficulty in getting the process accepted and really implemented by a field sales organization?

Answer

There are two parts to this type of situation. First, the field people have to be educated to understand the concept, the process, and their personal role in making things happen. I was just talking with another student who is involved with traveling to the field sites in her company and teaching the quality education system to them. She makes a trip to each one each month and conducts one of the ten courses. Then she calls people to see how they are getting along and encourages them to do their homework. A lot of good things are happening as a result.

The second part is that the company has to improve dramatically in what they send to the field. That is the ultimate component in having the process accepted. If everyone back at headquarters is running around wearing white robes and burning incense because of their conversion, but the same old stuff is still coming off the production lines, it is understandable that the sales force will not take it all seriously.

Relationships are what business is all about, and there are

only two that matter: the one with customers and the one with employees. Even here, in a business that specializes in quality improvement, we work as hard on employee relationships as we do on our customer interface. We have a monthly meeting called the "family council" where everyone comes together here in Winter Park. That meeting is videotaped so associates in other offices can see what is happening. In turn, they tape brief reports so we can see and hear them.

We have a company newspaper, a magazine, a rumor line—you call or write the corporate secretary and then the item is discussed at the next family council. We schedule executive availability so everyone can see the executives and be in touch with them. Doing all of this is simply a part of the reality of management.

Field people have to be looked at as part of the parent organization, not as someone who doesn't belong. It is a matter of concentration. When I took over supplier quality at Martin years ago, part of my organization consisted of 35 field quality engineers and inspectors. This was a group that no one loved and no one respected. Every time something came in from a supplier that had a problem, these people got blamed.

They tried hard but when two inspectors are placed in a plant with 5000 employees, they have little chance of catching everything and shouldn't be expected to work that way. So I decided that the reality had to change, as well as attitudes. I went out to visit these people and found they had low morale, low self esteem, and no idea of what was going on back at home base. Some of them hadn't been there for a long time and a few had never set foot in the place.

We brought them up to date on the status and set up an arrangement wherein they would return at least once a quarter for orientation and be given the opportunity to take some of their problems to the proper people. I found that it was necessary to establish a specific schedule inside the company for this. Otherwise people wouldn't pay much attention to them.

Inside the department at Martin there were many quality engineers who dealt with the field people and suppliers. They had a low opinion of the effectiveness and dedication of the field operation. So we fixed it up so that these people would go out and work in the supplier's plant when their field contact went on vacation. It only took one trip for each of them to change their ideas and ways.

After getting things working again, it became clear that we were not providing a proper learning base for these people. They didn't get the orientation others received. They weren't exposed to the regular training programs. Nobody loved them and they knew it. We overcame that. It took a lot of work and a lot of traveling, but they became very effective, the product quality improved, and everyone lived happily until someone came along who thought it was running so well that all the communication could be discontinued.

Question 59 _____

How do some of the most successful companies handle incorporating a far-flung sales force into the quality improvement process?

Answer

Someone flung those reps out there and someone communicates with them regularly in order to keep things going forward. That is the chain most companies rely on for quality education and support. The systems for dealing with sales people vary from company to company. But in every case they take whatever pains are necessary to reach everyone.

One company has an itinerant trainer who travels like a circuit preacher touching base with each in turn, using the quality education system. Another assembles people regularly by region and gives them their share of quality education along with the sales aspects of the meeting. The "quality awareness experience" can be conducted in four hours over a two-day period, permitting time for "soaking" and discussion. Films and written material can be supplied on a continuous basis.

But unless these people are trapped in a cabin up in the Yukon, they are in constant contact with the company in the normal course of doing business. That interface is the one in which they will participate and receive reinforcement after education.

If you think about it, most of what happens in the quality improvement process, as well as in the rest of running a business, is done by means other than face-to-face contact. In an organization large enough to have a far-flung sales force, there are a lot of folks who are not seen on a regular basis. Each entity can have its own team and its own information exchange under the umbrella of the company.

I worked in a 33-story building in Manhattan for years and saw very few of the several thousand employees who worked there. It wasn't until the annual Christmas party, held in a hotel, that I began to realize how many there were. Most of my contact was in off-site meetings. The home-office group might as well have been a "far-flung" force. There is often little routine contact between floors of a large business in a large building. We set up a series of seminars on quality, held them around the world, and brought people from all persuasions in to learn. In fact, ITT used to be said to stand for "I travel and talk." By that means, we reached the key people and then supplied them with material for those they affected. We didn't have the kinds of courses that are available now, but we were able to get the message across by including it in everything they received.

Many companies have communication systems that utilize in-house TV circuits, satellite transmission, and so forth. They can bring people together at their sites and have a broadcast. We are getting ready to do this with companies we serve because the requests for speeches exceed what I am able to handle. More sophisticated communications will permit doing speeches in the United States and Europe at the same time. It beats getting on airplanes and dealing with the whole airline experience.

Question 60 _____

Quality improvement efforts work well for manufacturing, but I would think that service organizations would experience totally different problems. Do you have to modify the approach for service companies?

Answer

Not really. The purpose is to teach how to deal with people and what those people produce, not how to deal with the product. Whether it is something in a box or an envelope or on a computer disk makes little difference. The purpose is not to help companies technically.

Now if I take IBM and ask what percentage of the people there are involved with actually building the product, the probable response would be about 50 percent. It is probably closer to 15 percent, but in either case that leaves a bunch of people doing what we are calling "service" jobs, those having to do with marketing, administration, service, sales, and such. There are an awful lot of suits worn at IBM. In General Motors probably 40 percent of the people touch the product. General Electric's most profitable operation is a loan agency.

The banks, brokerages, retail stores, and other service companies have huge factories in the back rooms where the product is tracked, customer transactions are taken care of, and the business of the business is assembled.

What I am saying is that the process of work in all companies is service. Individuals perform a service whether they are running a milling machine or a word processor. They are performing a service whether they are lifting a steel bar or a pack of reproduction fluid. And it is getting that service defined properly and then performed correctly that drives the quality improvement process.

We have to learn to measure the process of work as it moves along, not wait until a tangible product appears and then swarm all over it. That is where the perceived difference lies. It is possible to take a refrigerator and measure its conformance to requirements down to the last ten-thousandth of an inch. However, an insurance policy is much more complex if we are only going to sit and read it. The policy is an agreement between parties of something that is going to happen in the future should any of the described circumstances come about. If you think about it, a product fits the same definition. There is a description of what this toaster oven or that power saw will and will not do.

The reason people keep bringing up the difference between white-collar and blue-collar work areas is that the blue-collar ones have been measured since the beginning of the industrial revolution and controlled since Taylor moved the coal pile nearer to the furnace. White-collar areas just do what they want and no one really knows how many, how much, or how come. This process helps bring equality to the workplace, at least in terms of measurement and performance.

Question 61 _____

I'm part of a relatively small group in an entrepreneurial situation. Do you have any record with regard to implementing quality improvement from the beginning of a business as opposed to putting the system into a large corporation which is already manufacturing something and has a lot of momentum? If so, what sort of success have you had?

Answer

Our very company is a good example of what you are talking about because it is a new company. It is definitely an advantage to begin with a clean slate. The first employee I hired was my son—as accountant—and I sat him down and told him how the company would be run. We have done the same thing with every employee since. Everyone goes through management college and the quality education system. There are special inside classes. Everyone gets completely oriented. If you notice someone's desk, you will see that they have a little wooden block with a pen sticking out of it. That is what we call our ADEPT block. I personally present each new employee of the organization with it during the first week he or she is with the company.

The purpose of the block is to show what professionals are:

A is for accurate.

D is for discreet.

E is for enthusiastic.

P is for productive.

T is for thrifty.

If you need encouragement on what a small group can do, think back to the second world war. When the United States started to get ready for that war, they had a little tiny armed forces. In only a year or so they were ready to invade North Africa. The services were built up very quickly by training. Those of us who went to boot camp, or basic training, were taught the fundamentals very quickly. After the first day, there never was a bit of doubt in my mind about who was running the Navy and what I was supposed to do.

They sent me off to hospital corps school and taught me how to hold sick call, treat the injured, and be a field medic. I did that for two wars and everyone got along well. If something had happened to me, they could just yell for another third-class pharmacist's mate and the war would go right on. They took all of this seriously and so did we.

Now if a company is older, it can still change. Your body, for instance, changes millions of cells each day. So if you start a new pattern, like not smoking anymore, in a short time you have brand new cells that never heard of nicotine. Companies can be changed like that. I have never believed in this built-in "unchangeable culture" business. Everything can be changed for better or worse, and it will change by itself when the time comes. So it benefits management to cause proper changes rather than just waiting for events to run their course.

When I was at Martin as quality manager of the Pershing project, there came a time when the director decided that the quality division was getting a little cumbersome, so he wanted to reorient it. We designed a new set-up and I agreed to move over to being quality engineering manager. However, the general manager insisted that I go into supplier quality because that was our biggest problem.

This department had gone through several managers in the previous few years and was a constant source of irritation. The purchasing department was always screaming about how inefficient they were. The production people were always complaining about defective material they had to work with. And the suppliers were unhappy about the inconsistencies of judgment. One day something was okay and the next day it was not.

Stepping into this quagmire was not my idea of the rest I had in mind when I helped reorganize the operation. But the boss promised me anything I needed and authorized several more people for my staff. However, I put the requisitions in my drawer for a while.

The first thing I did was take my key people away for a weekend, which we paid for ourselves. I told them that their job was to get purchasing and the suppliers interested in conforming to the requirements, that we would live and/or die by the purchase order. Whatever was ordered was what we would accept.

We would be helpful in receiving inspection, but basically, we would look at things as they arrived and we would try to get the suppliers to send in defect-free material so we wouldn't have to do that. Purchasing would not be permitted into the receiving inspection area. At the time, they ran around there continually.

The team told me that everyone, including me, was unreasonable and that what we had to do was add 30 people and work three shifts.

On Monday I met with the senior purchasing people, along with our receiving supervisor and field quality manager. I took a big yellow pad and asked the purchasing people to tell me every problem they had with us. They yelled and screamed for an hour and I wrote everything down. The main problem was that everything got in late and didn't have clear requirements and we were a road block. Also we had no sense of urgency and were arbitrary. There also were dozens of little specific items.

The next week we met again and crossed off most of the items and added more. In the meantime, I sealed off receiving inspection and placed a desk for the expeditors to deal with rather than running all over the place shoving boxes in front of the inspectors and testers. I talked to the government quality people and told them that they were too easy and that anything they had a problem with was my problem too.

After about six meetings, the purchasing people ran out of things to complain about and asked if there could possibly be anything that we would like to mention. I suggested that we could work together to get suppliers to become serious about delivering defect-free according to the purchase order. They agreed to help hold some seminars. We also agreed to take a united stand to get engineering and manufacturing to take requirements seriously so the suppliers would know what was wanted. We agreed to meet each and every Tuesday morning and to lay out any problems that weren't getting solved.

All of a sudden the world began to quiet down. We held supplier meetings and told them we would help them to get zero defects. They changed rapidly. With no one to divide, the engineering and production people had to get their requirements straight. The quality field people were brought in and reoriented, those in-house learned to love those in the field. Inside six months the purchasing and quality functions were working hand in hand, the government loved us, the suppliers were happy. The only complaints we ever received came in the monthly meeting with my director and he would bring up one or two items that always turned out to be inaccurate. I couldn't figure this one until I substituted in his regular monthly golf game one Saturday. One of his buddies was an expeditor who would manage to slip in some problem he thought he was having every month. So I told him that the next time he had a big problem he should come to me and that if I ever got another one from him via the director, no one in supplier quality would ever deal with him again. Ever.

We never had any more difficulties. I gave back the unfilled personnel requisitions, wrote an article on the experience, and dozed off for a year. That is when I decided it was time to get out in the world and wound up at ITT.

Things can be whatever the leader wants, but it has to be spelled out. Every day is a brand new day. The problems that have been around for a long time don't have to be there.

Teach your people, explain things to them. Put them through the quality education they need to change their culture.

Question 62

At the store where I work, I am responsible for orienting all the new people who come to work for us. Most of them are between 16 and 18 years old. I would like to include a segment on quality in the orientation. How much detail do you think would be appropriate, given these circumstances?

Answer

That age group is a wonderful one to deal with. They are eager to please. They want to know what to do. I would suggest that you use something like the quality awareness experience course, which deals with situations they can understand. Facilitate it yourself. Then give them a tour of the place showing them specific examples of why it is so important to understand requirements and conform to them. Show them how to make inputs about suggestions, and let them know how important the customer is to the store.

The key to everything in this employer-employee relationship is providing a clear job description and measurement system that both parties understand. If you take time to do these things, you can count on your turnover rate being among the most favorable in the industry and your work force being the most effective.

McDonald's, for example, does a wonderful job with this age group. Every one of their places I have been shows the same bright faces, the same methodology. They have a TV and VCR in the back or downstairs and several brief films on how to do the various tasks. They train everyone, and if someone needs a reminder, they go down and look at it. Also, their commercials are role-modeling for the kids. They learn how to behave from watching themselves and their coworkers on TV.

We never outgrow our need for learning.

Question 63 _____

Is there anything unique about the application of the quality improvement process in a technologically oriented organization, say, for example, research and development?

Answer

The keystones of science are integrity and measurement. That is what the quality improvement process is all about, too. There are two special considerations for a research and development operation. First, it must be certain that what goes on in a given function is accurate and well-documented. Second, those who eventually are going to produce what research and development comes up with should do it exactly as planned.

In order for the second to happen, the first has to take place, and some communication has to take place. It is like doctors in a hospital. If they do not take the procedures seriously, then no one else will either. Being creative does not mean grabbing elements out of the blue and slapping them together in the blind hope that the result will be a viable new product or a new twist on an old product.

So the techies and other creative people should be setting an example and leading the way. The objective for a quality improvement consultant is to install in the rest of the company the same keystones that are supposed to exist in the scientific side.

The intent of the process is not to "invent it right the first time." The concern is not to create ideas for people. It is to foster integrity.

Question 64 _____

Prior to putting the quality process in a company, what kind of audit should be conducted to determine what basic managerial skills and organizations are already there? For example, are there adequate problem-solving skills in the company? Are the requirements for the process already there? Is the process defined?

Answer

It is necessary to spend time up front to get a better understanding of a company's situation and develop a preliminary price of nonconformance. The expression I prefer for this is "phase zero." However, doing an in-depth, detailed "capability" audit is not very useful. It seems like such a reasonable approach that many consulting firms do it as a matter of course. It is very expensive, though, and produces little.

First of all, few people in the company, if any, understand quality or can talk about it. This is not a criticism; they haven't been educated to it.

Once the education happens, then potential problems and functional inadequacies pop right to the surface. Everyone will recognize them. Audits just slow things down when a company is trying to begin improvement.

For instance, the "cost of quality" concept has been around for many years as a tool for auditing in the way you suggest. It began in General Electric as a way to compare the quality efficiency of different assembly lines.

When we started applying this concept to whole companies, figuring out how much it was costing them to do things wrong, management immediately thought this was a useful idea. However, those who were charged with putting it together fell into the crevice of indecision. They tried to pin down each nickel and couldn't decide if a situation should be labeled as a failure, prevention, or appraisal. As a result, hardly any companies were able to use this tool and didn't find out that they had desperate need of quality improvement.

Unfortunately, an extensive up-front audit is too often a way people postpone doing something truly useful.

Question 65 _____

What are the basics of successful troubleshooting?

Answer

First is finding which trouble needs shooting. When the early crusaders marched into Asia Minor on their way to rescue Jerusalem, they wiped out the first two cities they encountered. These happened to be the only Christian cities in the entire place. They were spring-loaded—one solution to every problem—and ended up eliminating their only allies.

When asked to look into a difficult situation in a company or operation one knows little about, it is advisable to start with two beliefs. First, those asking for help probably don't know much about it or they would be fixing it themselves. Second, there is someone there who does know and no one will listen to that person. If the troubleshooter can get loose from his or her host and wander around chatting with people to learn what their biggest problems are, a pattern will soon begin to reveal itself.

Usually they have been chasing the wrong problem. When the right one appears, someone has to be put in charge of getting things right and then preventing it from ever happening again.

I can think of a situation in which a randomly erratic electronic component defied the staff engineers for some time. It was all solved when they started assembling it in a clean room. The cause had been dirt in the contact area. This was pointed out to me by a line supervisor. She had been trying to get someone interested for a year. We didn't even speak the same language. She told me by pointing and blowing invisible dust.

I also remember a plant that consistently delivered units of the wrong color and fired three manufacturing managers for causing it. However, in the order department I learned that some of the sales people had been bypassing the order department and calling orders in to the floor. They didn't realize that some of the color code numbers had been changed.

One company's point-of-sale material never showed up on time because everything had to be done verbally. The reasoning behind this was that the boss of the marketing unit did not want anyone to make copies of any schedules in case the competition found out about them. Who was going to tell him that was dumb? I had no problem saying it.

Work is a process. This is essential to keep in mind, even if it means saying it until people get sick of it. What this means is that things go from here to there, to there, and then to someplace else following a pattern. If something is not coming out right, or nothing is coming out at all, there is a problem in the pattern. Either someone is not doing what is supposed to be done, or we don't know about something that needs to be done. The people in the operation know. They may not know exactly what, but they will know where, and you can figure out the rest.

Also, someone has to be responsible for everything—everything—or it won't get done. Set up task teams that will only exist until the problem is solved. Have them meet every day to go over the status for a few moments. Insist that action items be spelled out with responsibility and date, and give them the opportunity to report to management. Recognize them when the task is complete. If the problem doesn't show signs of being solved in an agreed amount of time, form a new team.

Question 66 _____

Are the quality improvement concepts and techniques you advocate generally applicable or will those of us who will go back and teach them be responsible for adapting them to our specific company's needs?

Answer

The quality education system is set up to let the instructor lead the students through an understanding of the concepts and an implementation of the tools. Films and viewgraph material, plus the workbook, cover all the material. Each session is half prepared material and half discussion of the application to it, all in that particular company. In the second half the students talk about what they are doing, and other people can come in to discuss progress.

But the quality education system itself as we have structured it is not something to fool around with arbitrarily. It is carefully designed to reach people in a way that will let them completely understand the full range of quality-related concepts and their role in making quality happen. The real name of the quality education system is "The Individual's Role." There is nothing in it about the 14-step process because that refers to a management effort. Everything is aimed at the individual working as a person and as part of a group or team.

Question 67

In your literature you spell out a lot of guidelines and methods. Do you find that some companies take the information and methods you provide and bureaucratize until they make the system inflexible?

Answer

Anything can be made inflexible and in the implementation of quality improvement, care must be taken to guide businesses away from actions that will slow them down in the long run. It is not something that happens very often but it does occur. When things take a nonproductive turn, it is necessary to intervene immediately and try to change the course. Usually everyone responds and alters course. After all, the main purpose of the process is to make businesses successful. Once in a while, however, someone will not heed advice, even when that person is paying for it.

I can think of three cases in which we have actually severed the relationship with a client. All of them returned later and are now doing well. The cause was usually that one person wanted to stamp the process with his or her own ideas and therefore made quality improvement difficult to accomplish. It is the worst kind of egotism to take something that has been 25 years in the creation and to modify it on the basis of an assumption of understanding and in an attempt to use it for personal ends. By changing it from something that is designed to be flexible, it is made inflexible.

The logic underlying the quality process is quite far-reaching; it is like a veil behind a veil behind a veil. What you are being taught here is the beginning. It is the readin', 'ritin', and 'rithmetic. Later comes the algebra and the literature, and after that comes the space mechanics.

Once a company has been in the process for a few years and has gained a lot of experience, it will find new things to add on, to emphasize. Quality education is a two-way street. We learn a lot from businesses that adopt this system, but to try to set up one's own alphabet in the first grade is probably not a good idea.

I have a favorite restaurant in Washington, D.C. It is French, and the specialty of the house is a soufflé—chocolate. It has to be ordered in advance and I'm careful to do so. When the soufflé arrives after the meal, the waiter practically has to struggle to hold it down on the plate, it is so light. Each and every time, it tastes wonderful; it is always exactly the same. There is not another like it anywhere.

Now, having checked all over the world—as I have—for a chocolate soufflé that works, should I then pour maple syrup over it in order to demonstrate my importance? Or should I use it to build a meal and thus enhance my life?

The other thing that can happen to cause inflexibility in the quality improvement process is for someone to decide that everyone is going to march along side by side. Each division is going to do step number three at the same time and so on. The result is that senior management ignores the whole thing because they think it is on a railroad schedule or something.

Nevertheless, almost everyone applies the system as it was designed to be done and gets wonderful results. There are a lot of case histories. Some of them have been filmed as Q-

21s, which stands for "quality in the 21st century," and many are documented progress reports that are kept in-house.

Question 68 _____

A number of quality consultants, who know our company is being trained in your quality improvement methods, have approached us saying we are going to need help later with the implementation phase. Do you think this is true and, if so, what is the nature of the help we will need?

Answer

Sometimes people assume things they shouldn't. When the Cunard line built the world's largest ocean liner in 1934, they decided to name it the Queen Victoria. So the chairman of Cunard made an appointment to see George V and told him, "Sir, we would like to have permission to name our new ship after England's greatest queen."

His highness smiled and said, "I am certain that Queen Mary will be delighted."

When people talk about helping trainees to implement the quality system we teach, they are usually off in two areas. First, they do not understand the concepts being taught. They have never been exposed to them. Our experience is that such people confuse the situation because they are not prevention-oriented, and they do most of their work in procedures that are encased in conventional wisdom. One of the main problems we have with them is that they base things around acceptable quality levels, for instance. It is a mistake to get involved in anything that sends out the message that you are not serious about zero defects.

Second, the quality management system that we put into place includes all the tools and support anyone could ever need. The list is extensive; there is a whole "tools division" that deals with statistical process control and other phases of statistics. There is software for statistical process control, cost of quality, process measurement, and other areas. Supplier quality seminars are conducted, as are corrective action workshops. On top of that, experienced professionals are available to work with trainees to whatever degree necessary.

Traditional quality control efforts do not relate to the implementation methods I advocate. Books full of procedures for people to follow are not very useful. If companies just try to meet needs that exist now, they will quickly learn what is missing and what needs to be clarified.

The Crosby Complete Quality Management System has been developed through years of international experience, through close examination of what works in the real world, and through what users have proven most applicable inside their operations. It is fully in place and it all fits together.

Question 69

In Quality Is Free, *step 8 of the 14 steps is called supervisory training. In* Quality Without Tears, *step 8 is called employee training. Why did you change that designation?*

Answer

When I first wrote the 14 steps early in my ITT days, the thought of reaching all employees in an education program was unrealistic. There were no VCR's, no videotapes, no cassette recorders, none of the things we take for granted now. People had to be taught face-to-face, with a film thrown in now and then. It was difficult to dig up projectors and very expensive to make films. I had to put the first zero-defects concept descriptions on reel to reel, and hardly anyone had the capability of playing them. The idea of training supervisors was that they would pass it on to their people.

By the time I wrote *Quality Without Tears,* we were using

all the systems that are so familiar now. So the old title of supervisory education was no longer operable or accurate.

There is nothing permanent about the 14 steps. They are like a road map; sometimes a better road map comes along. When I was a kid, we used to drive from Wheeling to Columbus, Ohio on U.S. Route 40. It was a two-lane road with occasional stretches of three lanes to permit some passing. Now there is an interstate that parallels Route 40. In fact, Route 40 is hard to find.

We do not make change without serious thought, but there is a time and a way to do it.

Question 70

What sorts of mechanisms are available to graduates who have completed their formal quality education but want to continue to learn from others?

Answer

There are several things along this line, available from a variety of sources. Of course, I can speak best with regard to what PCA provides to illustrate the kinds of alternatives you have. For example, PCA holds an alumni update conference every year around the end of May. It is limited to 250 graduates, and they usually represent 100 or so companies. Although these sessions are organized and managed by our staff, all of the input comes from the participating companies. They make speeches, conduct workshops, and so forth. It is a very useful activity.

In the regional offices (San Jose, California and Deerfield, Illinois) there are "user" conferences which are held regularly. The international operation has one each year, usually in London.

Then there is the *Let's Talk Quality* tape series. Discussion sessions are recorded and made available quarterly. This is probably the best way of keeping up with what students are currently thinking and what is going on in my mind. In addition to these, there is the *Quality Update* magazine, which appears regularly and contains case histories written by companies undergoing the quality improvement process. There are also several newsletters and the Q-21 videotapes.

The Q-21 tapes consist of interviews with those who are implementing the quality process or those affected by it. They are completely open and provide a great source of material for quality improvement teams, management, and others who want to see what else is going on. They are good for meetings also, such as meetings of quality improvement teams, executive committees, and professional and social organizations.

Once you are into the stream of resources available, there should be no difficulty staying abreast of the latest developments in the quality world.

Question 71 _____

How will we know know if our organizations are responding to all this change after starting the program?

Answer

The first sign is that people begin smiling more as the hassle begins to drop out of the organization. They begin to use different words. They begin to talk specifically about things. They refer to requirements. They stop looking at the stopgap, one-time solutions for solving problems, which compromises quality.

Quality has always been the most negotiable of the "schedule, cost, quality" triumvirate. Schedule has a specific time. When the 30th of October is past, there is no more, and you can't do anything about it. Either you made it or you didn't. Cost is the same. If the money runs out, the next step is to have brackets around your numbers in the reports, and people with brackets don't get invited to important meetings. So quality has always been the refuge. When the view of quality as a refuge begins to disappear, you know things are changing.

Another sign is that the quality improvement team will be very up and positive. They will be involved in all kinds of interesting things. They will start to collect success stories. They will begin to really understand the idea of "show and tell" and will begin to feel the need to arrange for people to present their success stories to senior management.

You will know. It is in the air.

Question 72

What is the probability of success when implementing the quality improvement process? Has anyone ever run a series of audits a year or two after the program is in place to measure its success?

Answer

The consulting business is one continual audit to make certain client businesses are successful. There is an obligation to help success happen and to take it very seriously. There are no failures in introducing quality improvement. No one ever gets worse and no one ever doesn't improve. Most of that is because of the extreme care taken in selecting businesses for quality improvement. It has to be certain that they are serious.

On the other hand, there are always companies that are not as successful as they could be. That is usually because they insist on doing some things their own way in order to stamp the process with their own mark. As I noted before, we work hard with them to convince them that Pa's chicken soup is not the same if you insist on adding raspberry sauce.

Those who execute the strategy developed with and for them and who work at the basics have remarkable success.

They improve dramatically and get a huge return on their investment. Once they have mastered the basics, they are offered a lot more things to work on and they can develop subsets of their own. Creativity flows freely when understanding is universal.

One business I know about, for example, came up with a great idea for getting this understanding locked in. They set up a "Quality Bowl" game, patterned after the TV College Bowl. Teams of graduates from quality education school competed, and the interest was amazing. Sessions of the contest—held on Saturdays, by the way—were standing room only. They put together questions from the quality educational system course and from trivia areas, such as what was my middle name. I found that one out when some young lady called me at home one Sunday evening to ask.

Success is a matter of return on effort. These concepts of quality improvement have been proven to work in any business, any country, any time frame. It takes a long time to really burn them into the brain so that they are automatically applied. For instance, in *Running Things* I talked about management's job being basically to establish an organization that can run whatever it is that needs to be run. It is more than just building one that manufactures widgets or sells loans. The concepts are broad and deep, generally derived and applicable, deeply rooted in psychological and sociological principles.

Not everyone recognizes success when they see it. I can remember the story of a fellow who was caught in a flood. As he stood on his front porch watching the waters rush by, a boat came along and offered him a ride.

"No thanks," he said. "The Lord will take care of me."

A bit later he was on the roof and another boat came by. Again offered a ride, he gave the same answer. In a short time he was standing on top of the chimney and this time a helicopter tried to save him. Same answer. He drowned.

Upon arriving at the pearly gates, dripping wet and sputtering, he asked why his prayers had not been answered.

The Lord looked at him with a puzzled expression and said, "I sent you two boats and a helicopter."

When checking progress, the good news is finding out what needs something done about it. The bad news is that something has to be done about it.

Question 73 _____

What sort of profitability improvement might be anticipated with the implementation of the quality process, say, as a percentage of sales?

Answer

A company can reduce its price of nonconformance to what amounts to 10 to 15 percent of sales rather rapidly, and much lower as time goes on and more causes of error are unearthed and prevented. That number raises the amount of operating income by a considerable degree. The expenses of the company are reduced, so more is available for other uses. Whether it is decided to put that into profit is up to the management.

There are documented cases of companies that have not had to raise prices for three years and yet have increased their profitability during that same time. Their competition, both domestic and international, cannot make that statement.

The true growth factor in quality improvement is the reduction of errors, such as in process improvement, and in getting more work out of people because they are only doing things once instead of several times. Together, all of these add up to continual improvement in every area, one of these being profitability.

But companies are like families. Let's say an artificial Christmas tree is purchased that eliminates the expense of buying one each year. What happens to that $50? Is it identified and put into savings? Or is it applied to having the family live better and safer? As waste is eliminated, as productivity grows, more things can be accomplished. The overall process of the company is cleansed of the snags and sandbars that impeded traffic in the past. The company runs better, much better.

As a result of all this, the company becomes more effective and more profitable. Above all, quality improvement is a long-range thing. It is not something to be looked at for a quick buck.

Quality Relationships: Toward a Quality Business Culture

Question 74 _____

In the beginning stages of a quality assurance program, you avoid having the traditional quality assurance people head up the program and the quality improvement team. Why is this? When and how do you bring the "experts" back into the program?

Answer

I always want the quality professionals to be an integral part of the improvement process. They should be in there from the beginning pushing and pulling to move it along. However, I do recommend that they not head it up. There are three reasons. First, it could look like just another quality program. By the way, this is as good a place as any to interject that I prefer the word "process" to the word "program," because programs go away. I like to think we are all seeking improvement over the long haul. Second, letting other people be in charge of the quality improvement team provides a greater opportunity for changing a lot of minds. Third, the quality professionals have not worked at learning a great deal about how management works, and consequently they usually aim the process exclusively at developing tools and at the lower levels of the organization.

Also it is part of strategy. I would like to see the quality profession get much more interested in prevention and in helping management cause that to happen throughout the organization. No other area of the company touches so much as quality does. Instead of being limited to the production areas—in both manufacturing and service companies—quality professionals should be branching out and infiltrating every part of the company. Instead of dealing so much with appraising products and services for acceptability, they should be continually measuring the whole process of the organiztion. Management really needs that kind of help.

There is a Quality Revolution going on and the quality people are missing it because they have their eyes on what goes out the back door, whether it's a product or a service.

They are concerned with waivers, deviations, and keeping track of problems. They should be concerned with how the customer is doing and what can be done to never ever have any problems or other nonconformances anywhere.

That is where the results come from, and that is where the future of the quality profession lies. If not theirs, someone's.

Question 75 _____

What, in your view, is the organization and function of the typical quality assurance department in a manufacturing plant, before and after quality improvement implementation? What is the proper placement of the process in the organization in the manufacturing plant? How can quality improvement be kept from becoming just a motivational program? Should it all be tied to cost reduction?

Answer

There are several questions here for the price of one. Let me answer them in reverse order. No, quality improvement should not be tied to cost reduction. Cost reduction is a byproduct of quality improvement, but people will not work hard to reduce costs unless their lives get easier along the way. Costs come right down, so it's not something to worry about.

To keep the quality improvement process from becoming a motivational program, it has to be implemented in a certain way. There is motivation in everything, but the quality improvement process has real substance that people bring right into their work practices. It is like the difference between a one-night stand and a lasting marriage.

The organization and operation of the quality assurance department needs to be well thought out in order to take advantage of the attitude changes that come about as a result of the improvement process. The goals of the department need to change from being a police force to becoming a wellness center. They need to learn how to examine the patient from stem to stern each day, to identify any processing steps that are not being performed properly, and to insist on permanent corrective action. They will do a lot of inspection and testing but very little of it for acceptability. Rather, they will be measuring the status of the organism.

In short, their operation must become prevention-oriented and lead the rest of the company by example, education, and results. Quality assurance is a very important part of the company. It should be a very well-trained and well-managed operation, but it should be kept small. As far as where it should report, there is no question of that. It must report on the same level as those it is measuring. Look where engineering, marketing, and manufacturing report. The quality executive should be on the same staff.

The purpose of this is not to have the "authority" to keep from being overruled. The quality improvement process blocks that out. What is necessary is for the quality manager to be part of managing the entity itself. Taking part in strategy sessions and becoming active in the day-to-day struggles of accomplishment are important. As part of this, there can be a tendency to get too supportive. It is therefore necessary to have a strong person in the job.

The quality department should put out a regular report—at least on a monthly basis—that talks about nonconformances and where they stand. The report should list items and detail exactly what happened, what specifically caused each to occur, and what is going to take place to assure that it will never happen again, anywhere. That is what "assurance" is all about. Not "superinspection" or test to catch things going out the door, not following up on a quality control department, but real prevention-aimed actions.

The quality department should be encouraging the quality

improvement team. It should be helping to set up "show and tell" sessions where groups and individuals come and witness to management about the things they are accomplishing. It should take an active role in the recognition activities of the company.

The quality department should lead the way in helping suppliers and customers learn to take requirements seriously and to undertand all the ones that now exist. The tradition of quality is very much in the hands of quality assurance.

The quality department should worry about the quality education and internal quality process of each functional department. They should make it easy for these groups to do for themselves, internally, what quality does for the company as a whole.

The way quality is managed today makes it easy for companies to downgrade or eliminate the function altogether. It is not viewed as very useful because it operates from concepts that have not been effective for 40 years. I know; I used to do that. Quality has to learn about general management so it can talk to everyone else. Quality professionals need to study speaking and writing so others can understand them.

There is a marvelous opportunity and a bright future for those who learn how to manage quality in a manner that makes it an obvious asset to the company rather than a "necessary evil."

Question 76 _____

How do reward and discipline fit into the quality improvement process?

Answer

I prefer to talk about appreciation and recognition and not deal in discipline. We are not trying to teach a dog to go get the paper and slippers. Dogs never really understand what you want them for, nor will they ever be able to comprehend why you raise that big white thing in front of your face for an hour each day. The dog will do the job once he gets the idea that it makes you happy and saves him a spanking.

So I do not encourage the carrot-and-stick approach. I always thought that was a dumb idea. People are thinking, caring beings, and they can tell whether you respect them or not. They work for appreciation and the sense of accomplishment they get from doing the job well. They do not work for money. They need it, and it is important, but money is a lousy motivator.

The Wall Street insiders were all people who had already gotten rich by being good at their trade. They didn't do the illegal things they did to get more money; they did it to know they were better at it than anyone else. Their motivation wasn't even good old-fashioned greed. It was the worst sort of pride. They didn't get what they considered proper recognition so they arranged their own. It didn't work out quite as they thought it would. Did you ever think what you could possibly do with a satchel full of hot money in small bills? One guy used it to pay his maid. He had $700,000 in a suitcase in his closet and it did him absolutely no good.

It is up to the management to arrange that the proper recognition be given to those who accomplish and the proper reeducation or reassignment be done with those who are having problems. The recognition step of the quality improvement process is one of its most important.

Again, to use our own company as an example, we have several awards, and our key one is what we call the Beacon Award. All of the associates vote and choose three of their coworkers—one executive, one professional, and one administrator—as the persons who are the best examples of quality performance in their respective jobs. Everyone is eligible except the chairperson. We present these awards during Thanksgiving week each year.

Now everyone else has Thanksgiving week in November. We do it in April. The quality improvement team runs the show, and it has become a tradition.

On Sunday they run ads in the paper or on the radio thanking the community for its support of their efforts.

Monday, each associate receives some form of thanks as he or she comes to work, and the members of the executive committee are usually presented with individual roses or some such thing.

Tuesday is supplier's day. Each supplier receives a certificate, and some get personal calls of appreciation.

Wednesday, there is a prayer breakfast behind the Vincent building to thank the Lord for His blessings.

Thursday is client day. Most receive thank you calls and a certificate. Those who are in class at the time get a special gift, usually at the open discussion.

Friday is family day. There is a child-oriented event behind the building. Last year it was hot dogs, magicians, and a ski show.

Saturday is the annual Thanksgiving Ball, a black-tie affair for all associates and their spouses or guests. We have the same big band that plays at the Symphony Ball. That is when we give out the Beacon Awards and have our pictures taken.

All of this costs very little, and it serves as a great reminder that we should be continuously thankful. Recognition is something that comes from being people-sensitive. We all need to work on it. Don't give people trinkets. A handshake, photo, or lunch will be much more appreciated.

Question 77 _____

In some of your earlier remarks, you mentioned that you once worked for a general manager who fired two people for not participating in the quality improvement effort. Would you address this negative side of recognition?

Answer

I wouldn't look at that as a negative. What happened was that when he took over, the general manager said that from now on the company was going to deliver defect-free products and services to its customers instead of what was currently being done, which was delivering whatever the customer would sign off on.

Two of his senior people thought that was impossible, impractical, and unnecessary. He tried to reason with them, but they felt very strongly about it. They also exercised a great deal of influence in the company since they had important jobs.

So he let them go off into the world and everyone else went about the task. The positive side of this was that everyone could now realize he was serious, and a negative influence was gone. I see no reason to have people about you whose views on things are so different than yours that they will hurt the organization. Those were two people out of an employee pool of 10,000.

However, I have not seen this kind of thing happen very often. Usually people say, "It's about time!" and get happily on with the job.

Question 78 _____

What about the other side of the coin regarding that last question? What if the rank and file are in favor of improving the quality of their output but upper management isn't? Why might upper management be against entering a quality program?

Answer

There are lots of cases all about us where people refuse to do things that are good for them and will make their lives easier, happier, and longer. They have a different personal agenda, and it is sometimes difficult to determine exactly why they have it.

As chairman of the Heart Association fund drive and an active member of the Hospital Wellness Center, I constantly see cases in which people know better but won't do anything.

Certainly no literate person can doubt that smoking, being overweight, having high cholesterol, and doing no exercise are bad for you one at a time, and a tragedy waiting to happen when existing all together in one person. Yet I know personally at least six people in that situation. What's more, they worry about it too, which adds to the time bomb that is ticking away in their lifestyle.

The big problem usually is that they just do not know what to do or how to go about it. There is a lot of confusing information on what is right and what is not. For that reason they should go to a hospital-managed wellness center. There, they can get the scientific system of improvement, their family can receive counseling, and the whole culture can change. Often the family, while encouraging the potential victim to reform, will actually provide the wrong message by the way they run their lives.

The situation as you describe it, with the initiative for quality improvement coming from the lower levels of the company, would probably best be handled by a party outside the existing organization. A good quality consulting operation should work as a kind of "corporate wellness center" for quality. The educational systems should be specially designed to fit every person involved. It should provide the necessary tools for implementing quality improvement. It should always stand prepared to provide counseling in quality matters. Every attempt should be made to be completely objective.

With such professional "wellness" counseling, top management, which has just not known what to do in the past and thus has been inactive, is enabled to suddenly step out in front of the parade and say "follow me."

Question 79

Have there been any studies done on personnel turnover in connection with implementation of the quality improvement process?

Answer

As companies get involved in quality improvement efforts, their employees become happier, so in every case I have ever heard, the turnover drops. Proper understanding and proper communication reduce hassle. Most feelings of frustration in any relationship come from not being heard.

I can remember as a kid that when children would speak up or ask questions, the adults were inclined to downplay our involvement. They would say things like, "Children should be seen and not heard." That sort of behavior is a result of custom and environment.

When we hired our personnel director here, I told him that his job was to see that the employees get a fair deal and that the company gets a fair deal. He was not supposed to protect the company from its employees, which is what personnel departments usually think they are supposed to do.

Not having a common understanding of quality puts more pain into an organization than anything else I have ever known. I once worked on a project team as a line quality manager. There was a project director and then there were managers for every function: engineering, planning, manufacturing, marketing, purchasing, and all the other "ings." The nine or so of us ran the project. I reported to the quality director for "how" and "who" and to the project director for "what" and "when." A lot of people have trouble with this kind of matrix arrangement, but I have always liked it.

At any rate, the members of this team became very close to each other and, of course, we all wanted to succeed. Every Friday we would have an all-morning team meeting to go over the status of the project in exquisite detail. At each meeting, quality would become a big subject because of something that didn't work or wasn't quite right or was delayed. Everyone would look at me. They knew that all I had to do was say the thing was okay, to make a judgment on it, and away would go the problem.

They felt that quality was a matter of deciding how the most recent nonconformance fit into the overall plan of the project. Did it affect "form, fit, or function?" Could we live with it? Would the reliability of the system be adversely affected? Well, every little thing in itself didn't necessarily mean there would be a permanent disability, so on a case-by-case basis it became very hard to say no.

When you are part of a team and you have the power of saying yes or no, it becomes hard to take the straight path. Others just don't understand that the result of agreeing to this small exception here and that medium-sized compromise there and this larger but understandable problem somewhere else is a general deterioration of quality. Requirements are being adjusted to meet the pressures of cost and schedule. It was very hard being a quality manager.

This reflects itself all the way down to the people who find that things can be accepted, no matter what their condition, if management wants it to be so. They lose faith in the integrity of the entire operation, become frustrated, and all the things that grow out of bad morale begin, including turnover among the better people.

When everyone understands and agrees that requirements can be offically changed if they are found to be incorrect, unnecessary, or obsolete, then they quit putting pressure on the poor soul who is running quality. *Perform in accordance with the requirement, or change it offically to what the company and the customer really need.*

I used to hate those Fridays when I worked at Martin. So when I went to ITT, I was determined that this was not going to be the case. The first time I went to Mr. Geneen's monthly general management meeting, which lasted three days from 10:00 a.m. to 1:00 p.m., I waited for the first mention of "quality problem." It occurred in the first hour. As soon as someone said those words, 79 heads turned and looked at me.

"Don't look at me," I said. "I'm the one who told you about the problem, which is rooted in the design. I don't do design, marketing, purchasing, manufacturing, or any of those other things. I measure, report, and try to get everyone interested in preventing problems. So do not expect me to take on everything that was done wrong and wave my magic wand over it. If you don't want a requirement met, then don't establish it."

There was a silence. Then Mr. Geneen asked me if it was all right to continue now that we understood each other. I nodded and we went on. I overreacted a little, perhaps, but it worked because I never had that "quality-problem" problem again.

That is one of the things I was concerned about with the NASA situation, that they would dump the whole problem off on the quality people. Unfortunately, most quality professionals think they are responsible for quality and so become unwitting accomplices to their own frustrations.

Question 80

How can you ask people to actively work to reduce the cost of nonconformance when, in fact, they may be thinking that by doing so they are going to eliminate their own jobs?

Answer

First, the main reason for doing quality improvement is not cost elimination, it is to satisfy customers and get the company in better shape. Cost reduction is of major interest also, but it should be looked at as something that saves jobs, not something that eliminates them. There are a lot of expenses that can be eliminated when everyone works together. Reduced expenses are a good component of job security.

Worry about quality improvement eliminating jobs is not something I have seen as a problem. It just never comes up. People are in touch with real life. They know that a company has to continually improve or it dies. They know that costs have to keep coming down. They know that quality has to be managed. You will find them very grateful that leadership is being applied in these areas.

There is a tremendous opportunity out there in the world for a company and its employees. I was in China recently. They have one billion people and 100 privately owned automobiles. They have few roads, telephones, wristwatches, clothes, houses, few everything. There is a lot of work for the right companies. But it is going to take a while. China is larger in size than the United States. It has one time zone. Think about it. The sun is overhead in the western part at 3:00 p.m. That is just one sign of how many things they are going to require over the years.

Your employees will see more hope than fear in a well-planned, well-managed, quality improvement process.

Question 81 _____

Up until now, the charts and graphs we use where I work have been set up so that progress is associated with increases and lines that go up. Many of the charts and graphs you use associate progress and improvement with decrease, for example, zero defects and lowering the cost of nonconformance. I know it's just a matter of perspective, but do you think this could have a negative impact on the people who are seeing these graphs?

Answer

By going up and going down you mean the direction the line on the chart goes? Well, you can have it any way you wish. I don't have any particular preference one way or another.

It is best to follow the path that comes most naturally. If you are tracking something in which "more is better," like revenues, then the chart will naturally have the high numbers on the top and the line will make you happy when it rises. If you are tracking weight and want to lose, then the high numbers will still be on the top, but the happy line will be going down.

The key is to make very clear what the chart is measuring. Defects want to drop. Suggestions want to rise. When everyone understands what they are looking at, there is no negative.

The failsafe way to manage this is to have the people who are affected by the chart decide how it should be presented, where it will be hung, and what it all means.

Measurement is something we live with continually. On my body at this time there are several instances of measurement: my watch, a pocket calendar, the sizes of the various articles of clothing, my five-year pin, and so forth. My body, inside and outside, filled up two pages of measurements after the wellness center got through with me. Blood alone took up a half page.

But if we don't measure, then we cannot communicate. If you look at the weather report anywhere in the world, the numbers tell the story even if the words are not understandable. When driving, most people follow a numbered road system and keep track of miles traveled, gasoline, air in the tires, money spent, the time of day, the number of kids in the back seat—a whole bunch of stuff.

So it's best to look at measurement as a normal part of operations, something to be taken as part of the scene. Don't make a big deal of it, it is normal.

Where companies get into trouble with measurement is in how they use it. If people are fired or made to feel ashamed when something goes over a line, or if management itself doesn't understand the chart—which is the biggest problem of statistical process control—then measurement comes to be seen as a negative thing. But any good thing can be turned negative. Some people even use Santa Claus to scare kids.

Question 82 _____

What do you do in a situation in which expected performance standards exceed the minimum stated requirements for a job or task?

Answer

Let's see if I can come up with a "for instance" to make sure I understand your point. You mean that if I agree to mine 16 tons of number-nine coal and deliver it to the stock room, I am going to find out that you really expected 17 tons?

Well, if that's what you mean, that's a good way to get into trouble—having requirements that are not real. I remember in high school when the assistant coach who led calisthenics believed that everyone goofed off while doing them. Therefore, he had us do about one-third more than necessary. As a result we all goofed off, which reaffirmed his original analysis. He'll go through life never knowing that he caused his own low opinion of player ethics.

We have to be honest with people if we want them to be honest with us. So we agree on what they are to do. That is what we expect, it is what they expect, and it is what they do. It comes out properly that way. If they cannot trust what has been agreed on, then the word that describes the result is *anarchy.*

Question 83 _____

How does the quality process impact innovation in job methods and also in the manufacturing process?

Answer

Innovation in a manufacturing organization comes from the opportunity for innovative things to be implemented. One of the big problems the communist-bloc nations face is that everything is so controlled, regulated, and suppressed that individuals stop having ideas—or at least they do not share them with others. Suggestions for improving can be looked on as a criticism of the management or the professionals responsible for that activity. That can get you in trouble.

Many U.S. companies create the same environment. They defend the status quo as it is now and do not make any serious effort to continually improve. They build walls around the specialty groups, such as manufacturing engineering, and keep everyone out. The result is that innovation grinds along very slowly.

When a quality improvement process is installed, everything opens up. People don't have to be defensive and look upon the need for improvement as a criticism. Everyone is involved and will benefit from innovation and the improvement it causes.

Another side of this same issue is that creative people become discouraged when the things they create are not implemented properly. If a company does not have a policy of doing things right, then the innovator is always going to be disappointed.

I remember back at Bendix, when working on missile systems as a reliability engineer, I wanted to get everyone more interested in reliability so I ordered some little stickers to place on the telephones. They said, "Does it effect reliability?"

When the stickers arrived, they read, "Does it *a*ffect reliability?" I went to see the purchasing agent, who told me that my copy was wrong and that he had corrected it to save me embarrassment.

I explained that I was looking for an action word. "The player effects a touchdown, and that affects the score." I said by way of example. "I want to say, 'Does it cause reliability?' not 'Does it alter reliability?'"

"People won't understand that," he replied. "It will be better this way."

I asked how that got to be his job and we had a disagreeable discussion. I took the box of stickers to my boss. He then took them to his boss, and the final resolution was their decision that I was using the word incorrectly. They stayed with the word *affect*. It doesn't take much of that kind of foolishness to discourage an innovator.

The error cause removal part of the quality process will produce input from almost everyone in an organization. Traditional suggestion programs get 5 percent response or so. Only the toughest will face up to a multicopy form and a committee review.

A well-run error cause removal effort will produce several inputs per person each year.

Question 84 _____

Sometimes customers are not really sure what quality requirements are because they are focusing on performance specifications rather than objective product specifications. If performance standards are more flexible, the customer may accept something that deviates from product requirements. The problem is with employees, workers, lab people, etc., who see that nonconforming products are being shipped. How is this kind of problem to be dealt with?

Answer

The answer is that we have to be honest with ourselves, our employees, and our customers. Talking customers into accepting what they didn't order is an old and formerly noble profession. However, now things have changed and we don't do that any more because we know that they are going to get another supplier if we do not follow the policy we have of meeting their requirements. But who needs a supplier who constantly provides surprises and not-quite-conforming material? There are people all over the world who would love to take that contract.

We need to concentrate on establishing clear requirements and then meeting them routinely, not on explaining how the wrong mixture is really okay.

If nonconforming material is being shipped, the people are right to feel the company is not interested in quality. And zero points are made with customers when they are always being asked for deviations. Tear up the deviation procedure. Get rid of the forms that are used to make one out.

Question 85 _____

Do you think it is possible to improve and build customer service and still maintain a personal relationship with customers?

Answer

The quality improvement process, at least as we have developed and teach it, is designed to bring employees and customers much closer together by providing an understandable concept and positive channels for communication.

The typical field service person usually finds that most of the job involves explaining to the customers that the company really wants them to succeed, no matter how things may appear. Field people would love to have the feeling that the entire company is behind them.

One big corporation I know of has recently completely changed its field service concepts. They have removed the very thick book which lists all the things the field service person is not permitted to do or must have special arrangments to accomplish. They have replaced it with a one-sentence policy that says, "Whatever the customer wants to do, that is what we want to do." The stuffed-shirt days are gone for that company.

The "personal feeling" that exists now in an unfortunate number of cases is a situation in which the "lone ranger" out there in the field is trying to protect the customer against what the insensitive people back at the ranch are doing. When the quality improvement process is installed, cooperation and communication become the actions to live by. Then everyone is more personal.

Question 86 _____

Do you have any successful strategies for how to deal with a supplier who provides a unique product that can't be obtained elsewhere (or from anyone else)?

Answer

I'll tell you about a way to cope with this situation, but first let me stress the importance of not leaping too quickly to the conclusion that a sole-source supplier is necessarily wrong. Many of you are married, right? How many wives or husbands do you have? I rest my case.

The customer and the supplier have to have a relationship based on mutual need. No business ever becomes so successful that it can do without customers, and very few can get along without suppliers. Even bank robbers need somebody to stick up.

When a situation exists in which a supplier has a lock on a particular product or service, then it is necessary to work on the relationship—that mutual need. Make the supplier part of your family, and at the same time let it be known that it is always possible to design them out of the system, if the need arises. After all, there are no marketing advantages that last very long. Something always comes along to serve as a replacement or substitute, or to eliminate the need entirely. Your supplier knows that too.

Getting requirements clear between supplier and purchaser is one of the most important parts of the quality improvement process. It is particularly true in a sole-source arrangement. Think of it. Most of the suppliers we have in our personal lives are sole source: the family physician, dentist, banker, broker, grocer, gas station, commuter train, hairdresser. We can change any of these if we want, but we prefer to try and get one trained to do things the way we want them done. It is the same in business.

Also, it helps as part of the relationship between yourself and the supplier to be able to identify some specific cases in which people who had the market "locked up" were overtaken by others. I think if you talk to the IBM people, you will find that they have learned to not take anything for granted. The Big-Three automakers had 99 percent of the U.S. market only 20 years ago. The major TV networks had 100 percent of the U.S. market 20 years ago. In each case they lost market share because they became complacent and arrogant. They developed a "works in the drawer" philosophy of service rather than a reliability philosophy.

We had some people attending classes here a while back— the 44 top managers of a product division of a very large corporation. These people had 80 percent of the market in the United States. They were dragged in here by the division president. They arrived mad and they stayed that way for two and half days.

I went over and talked to their classes to no avail. They were walking around the room while I talked. Now not everyone agrees with me, but hardly anyone ever says I'm not interesting, even though they may think I'm misguided. Finally I asked these managers why they had come at all and said we'd never had such a group. They said the boss had insisted but they knew all along that they didn't need to improve. We offered to let them go home and said we would teach the class to an empty room. But they stayed because we wouldn't give them the money back.

Today, that group's share of the market has dropped to 45 percent, and every one of those people has left the corporation, every single one. There are enough stories of this sort out there to keep anybody from becoming complacent.

Question 87 _____

My company is a manufacturer of components. One of our suppliers happens also to be the buyer—our customer for the component—and we're having a problem getting them to supply us with a zero-defects part. What, in your estimation, might be the cause of such a problem, and what can we do to remedy this situation?

Answer

I would think that this supplier-buyer would be very interested in helping you, but probably there are two different parts of their organization involved. Most companies keep their purchasing and marketing organizations separate so they cannot put any pressure on suppliers or customers. They do not want to get accused of violating the reciprocity laws.

So I would suggest treating them as two separate areas. Insist that the part of the company supplying you with components comply with the purchase order and deliver what has been promised. Don't even acknowledge that another part of their organization is a customer. Don't give them any special considerations.

Many companies have internal suppliers, and they are always having trouble with them. This is called vertical integration. From our perspective as quality consultants, we see a lot of cases in which the diverse branches of companies get caught up negotiating yields and basically making a lot of trouble for each other. Management adds to this when it gives these "captive suppliers" profit goals.

It is one of our major objectives to help companies rearrange their thinking on this. Supplier groups have to concentrate on making their customers successful.

Question 88

In dealing with suppliers, how do you break through the bureaucratic maze to get to the right decision maker in the supplier's organization?

Answer

Dealing with suppliers is a learned process, and there is enough to learn on this topic to fill a two-day workshop at the Quality College. First is to have the right level in your company contact the right level in the supplier company for the purpose of letting them know you are serious about quality improvement.

Second is to explain to the suppliers, usually in an open-meeting environment, that their future depends on the two of you agreeing to requirements and then having defect-free products delivered on time.

Third is to listen to the suppliers in order to help them do what is wanted, which may require some changes in the purchasing of material or some other related practice.

Fourth is to not blindly direct them to do something like "install statistics" in order to accomplish quality improvement. Insist that they do what is proper for them, and ask them to show evidence of what is happening.

The main deal in any relationship is getting the requirements clear. Here's an example. Every year in central Florida, there is what we call the "Shark Shootout." It was created by Greg Norman, the golfer, in order to benefit the Arnold Palmer Children's Hospital. Norman talked Jack Nicklaus and Ray Floyd into joining him and Arnold in an exhibition. This year Fuzzy Zoeller was there instead of Floyd, who had schedule problems.

Well, I made a donation to the event and so was invited to a private dinner at Norman's home, where I got to stand and stare at the British Open trophy. As the dinner went on, one of the arrangers came over and asked if I would like to be a caddy for the shootout. I ran the thought through my mind, pictured myself strolling along inside the ropes offering advice to one of the pros. So I agreed and was signed up to caddy for Jack Nicklaus.

The next day there was a brunch in a tent and the golfers answered questions for the sponsors. Outside, about 2000 other people were gathering to watch the match. After the brunch everyone headed for the practice range, and I went to the first tee since the range was crowded. Then the golfers came to the putting green, which was where I was standing. As they putted around, a fellow came up and asked if I was Phil Crosby. When I nodded, he plunked Jack's 100-pound bag down in front of me and walked off.

At that moment Jack came up and handed me three balls and his putter, instructing me that the cover should be placed on the putter before being put back in the bag and that the balls should be rotated. I found myself—I who can speak before 10,000 people without a moment's hesitation—unable to tell this legend standing before me that there had been a terrible mistake. He smiled, patted me on the shoulder, and bulled his way through the crowd to the tee.

I picked up the bag, groaned, and followed. The real situation was beginning to dawn on me—the requirements—and I looked about for a strong young person who would like to make $50, or even $100. To make a long story short, my son-in-law came and took the bag after about 150 yards. We stuck with Jack for six holes, Nick carrying, me passing clubs and wiping balls, until a real caddy appeared.

This year the idea was not revived. In any case, in the future I will try to communicate more thoroughly on the requirements before agreeing to a task.

Question 89

Could you address the points of conflict when quality improvement education comes into play with corporate politics?

Answer

You can't teach executives inside their own company. They just will not take any of their peers seriously, or anyone else for that matter. It requires a credible source outside the operation, someone who does not want their job or any of their power. When they see that everyone is learning the same thing, that it is all for the common good, that no one person is going to get special credit for it, then they will go along and make it happen.

It is also necessary for those who direct the process to be careful in making certain that the executives are properly recognized when there are bouquets to pass out.

That is the rationale behind the creation of something like our executive college. You need a place for "brain transplants" that take away the need for politics in quality. Here is some background to show how the ideas evolved.

At ITT, I had around 500 quality managers involved in 87 divisions. That meant at least 500 general managers, 87 division presidents, plus group executives, corporate executives, and a bunch of other important people. They all had to be dealt with or nothing was going to happen.

The first thing that was necessary was to make certain the chairman thought what I was doing was the right thing, which was not difficult since he instinctively understood what zero defects meant in terms of earnings and lack of hassle. (If you read the dedication in *Quality Is Free*, you will see that Mr. Geneen is the one who thought up the title.)

Now, having the big boss on your side only provides protection as long as what you do is useful. This was not difficult, since our staff spent its time helping operations rather than criticizing them. It is so easy to find things wrong and report them that most staffs fall right into that situation. We took special care not to. Again, there is a whole section on that in *Quality Is Free*. Getting the job done correctly and having everyone appreciate you at the same time is an art in itself.

Recognition of these senior people was based around the Ring of Quality, the only corporatewide award ITT had. I started off by stating that the office of the chairman and I were not eligible for this recognition. Each of the recipients was given an imbedded ring to set on his or her desk. That way, visitors would see it and ask about it.

The ring itself was—and still is—a peer award in which nominations came in from all over the corporation, and about 40 were given out each year at a formal dinner. However, rings could also be given out for significant achievement to be recognized by the Executive Quality Council. Thus each senior executive received a ring at the time the operation was in complete compliance with all the goals of the quality policy. When the councils were working, when the 14-step process was being installed, when the senior executive was a living witness to quality, then a ring was presented and everyone knew about it.

Everything was positive. There was no need for trying to manipulate anyone or do anything except what was up front and positive. There are no politics in quality unless someone is threatened with failure and public exposure. It is vital that the quality department feel that it is part of the overall process, otherwise it will be tempted to be defensive about the whole thing.

That is why it is necessary to declare a general amnesty as soon as the improvement process begins. What has happened until that time is declared to be no one's fault. No one is going to come under attack. The goal is to attack problems for the benefit of everyone.

Question 90

In three weeks I have to present a quality program to my board of directors. They are the kind of people who don't want to hear about problems and I think this in itself is a problem. Do you have any suggestions for how I can approach them?

Answer

It is necessary to show them that they are a part of the situation and at the same time not get them upset with you. I would suggest you try to get your hands on a copy of a BBC film called *The Quality Man*. This film was made on a golf course in Scotland and has the philosophy of quality management explained in terms your board will be able to understand. It is very well done.

It starts off with the narrator saying that it took 12 years before he realized that management was supposed to be helping to achieve quality. He had been assuming managers were a punishment from on high. By the way, I should tell you that I was the narrator for that film.

After you have thus disarmed them, you explain the process—not program—that you have been learning about. Tell them what you are going to do, what support they will need to provide, and how they are an integral part of the witnessing aspect of the process. Don't ask if you can do it; assume that direction. Then arrange to include them in a "show and tell" session several times a year in which groups who have achieved something assemble and show their achievements to the senior executives.

The board has to understand that this is not a problem that can be solved by procedures or regulations or programs. It involves having to change the way the company is run. Everyone has to understand each other the same way.

There are quality conferences going on all over the country. The American Society for Quality Control just had its annual meeting, as it has for almost 50 years. Yet at each of these conferences there is no agreement on what quality is. They have little idea of what they are up against because they are trying to solve a people problem with techniques.

I got into a lot of trouble with my colleagues on this subject because the society just put out a "Quality Manifesto" which they asked all the past presidents to sign. It started out with the words "High quality is…" and exhorted everyone to achieve it. I could not sign it. They still seem to feel that quality is "goodness." But goodness is in the eye of the beholder, or at least is nonspecific. It is very difficult to deal with a subject when people cannot agree on the meaning of key words.

The opposite was very noticeable at a recent alumni conference of graduates of the Quality College. There everyone dealt with quality the same way. They shared a common understanding of what was necessary, and they spent the whole time talking about implementation. They were able to share ideas with each other in a useful way.

If you get your board and the other senior executives to use a common language of quality and to recognize their personal role in making it happen, everything will come out just the way you want it.

Question 91 _____

Corporate culture has traditionally been one in which people who fight fires are visible and get ahead. The quality culture, with its emphasis on doing things right the first time, is a different culture. How can corporations be successful in identifying the "do-it-right-the-first-timers" and in moving them ahead when the management that is in power got there by their fire-fighting skills?

Answer

Usually those who have to scramble to get to the top are well aware that there is a better way. They even yearn for it to exist. They have learned that fires only get banked, not extinguished. It is one thing to do it yourself and cope with a world where people do not do things properly. It is quite another to supply direction and delegation to those who are not doing things correctly and then suffer the frustrations.

So attitudes and needs change. The power struggle you suggest is eliminated even before the process begins. Remember, these fire fighters are the people who must approve getting involved in the quality improvement process in the first place. They are the ones who want to change and want that culture to be altered.

Be sure you are watching closely. It is not always apparent when a culture undergoes a transformation. More often than not, the need for a culture change occurs before those in charge of the operation know it is needed.

Processes, such as building automobiles, mixing chemicals, writing insurance policies, and so forth, are conceptually set up to make something in a standard way. Every operation is planned down to the last detail. When a nonconformance arises—something doesn't go together right—common practice is to repair the flawed product after the process is over. So auto plants have parking lots full of planned rework, chemical plants do a lot of "blending," and paperwork operations allocate a large portion of their time for checking and correcting. Paperwork companies patch up after their work is done. Computers help this happen.

I have been in assembly operations where they have bypassed the problem by learning to make the subassemblies wrong enough to fit! People become very inventive. But they have always fixed things for the life of the product, which may be years.

Once they become involved with the concepts of quality improvement, they realize that it is a wonderful idea to learn how to do things right. So when an incompatibility appears or an error repeats itself, they have learned to take corrective action to eliminate the problem. The idea becomes paramount to produce within the assembly process, not after it is all over.

If you think of a riverbed with rocks and snags and sandbars all around and a boat trying to flow down it on schedule, you can get the idea of how a company culture works. Management has three choices: (1) to learn to live with the delay and inconsistency that comes from the boat getting hung up on a regular basis; (2) to spend a lot of money and effort to put more water in the stream, which increases overhead and reduces profitability; (3) to clean out the river bottom by removing the obstacles to defect-free operation.

Removing obstacles makes a dramatic change. All the "after the ball" operations disappear. The enormous rework activities go. Consequently, the customers receive what they ordered and are much happier about it. And since everyone is trying to reduce the number of suppliers they have, the company doing things right has an increased chance of being one of them.

Now changing this way of doing business is not something that gets done inside the plant or office. Usually the people there cannot make it happen by themselves, even though they would like to. They have to have permission from their leadership. And this leadership, which has routinely exploded any time the process stops, has to convince the rank and file that they are serious about changing. They have to really witness in a practical, realistic way.

Once this begins the people can do things that they have known should be done. To use the automobile plant again as an example, the painters always wanted to paint cars of the same color in order. But because of the way the computer was programmed, they ended up doing them randomly. One would be blue, the next red, the next green, and so forth. That meant the paint lines had to be cleaned between cars, paint flushed away, and the whole job was just not as efficient as it could be.

Now, after some quality education, they paint all the blue cars for the shift, then go through each color as a group—a lot less hassle, a lot less waste, a lot better paint job. They also do a much better job of keeping the paint lines clean. The suppliers have learned how to produce dirt-free paint, and the paint areas are being changed in order to be dust-free. The change is dramatic.

In many areas the "tag man," the person who would take someone's job during that person's break, has been eliminated. The whole plant shuts down and everyone takes their break at the same time instead of individually. The reason for this is that a great deal of the rework was caused by this person doing a dozen different jobs. But this wasn't permitted before because no one wanted to "stop the line," which is the kind of simpleminded, imbedded culture that companies tend to lapse into over the years. This practice probably started years ago when one senior person erupted over some trivial line-stopping incident.

As organizations learn to keep their process up to date and

to operate it with the idea of delivering defect-free products and services to their customers, on time, then they will always be improving.

Don't adapt the quality improvement process to the culture, change the culture to conform to what is best. Learn from the past, but don't live in it.

Question 92 _____

We have a difficult problem where I work; you may have encountered it. How would you advise handling hassles and conflicts with one's boss?

Answer

The difference between discussing and arguing is whether the participants are using facts or opinions. On the subject of quality, it is a good idea to ship the boss off to management college in order to gain a common language in that area.

There are some people who *like* to hassle others, either because they think that is the way to manage or because the other person brings that out in them. It may also be that they do not realize they are making things difficult for others. There are a lot of difficult people who actually think they are charming. There are a couple of things you can do in order to reduce the hassle to zero and build the relationship with your boss to your advantage.

First, don't argue about things that are factual. If the discussion is around how many people there are in Florida, don't guess, go look it up. Just do not respond to opinions and guesses when they are put forth as facts.

Second, write down what you think you are supposed to be working on and ask for confirmation. Then stick to that, making modifications when other things come up. Supply the boss with the list, and keep a regular status coming in. All of this will take only a small portion of your time. Spend the rest doing the things that are really necessary.

Third, be pleasant and cooperative, but don't let anyone beat up on you. Life's too short for that.

Once, when I was a quality supervisor, the chief engineer of the company called me to his office, which was an area I did not get to visit very often. He was all upset because we had rejected some castings and would not agree to use them because they had porous areas. He really laid me out about this and told me how much trouble it was causing him and the rest of the company. Then he put forth some lightly concealed threats about the terrible things that were going to happen to me if I didn't start being more cooperative with engineering.

All this shook me up. I wasn't used to such treatment and I think he intimidated me. However, I went back and did what I was going to do anyway, which was to scrap the castings.

A few weeks later he sent for me again, but I replied that I would not be able to come. I went to the quality director, who was in the peer group of the chief engineer, and explained why. When the chief engineer came to complain about me to the quality director, the answer he received was, "Phil doesn't want to come to your office and talk to you because you are rude and nasty. He feels it is not necessary to do business that way."

To my surprise, and to the surprise of the entire organization, the chief engineer came down to my office on the factory floor, apologized to me, and we had a nice chat for about 30 minutes. We did well after that, although his integrity varied with the schedule as always.

There is a chapter in *Running Things* on performance appraisal which relates to all this. When people become more trouble than their contribution, it is time to do something about them. But in most "people relationships," an honest, open chat about specific items will resolve most conflicts.

Question 93 _____

How do you deal with the skeptic who is in a key position in the company?

Answer

The true skeptics are usually those who only want more than an emotional "feel" in order to be convinced. They want to know what is in it for them, and they want to be sure that it is going to be good for the company. They want to be certain that any given change is not a fad, that it is based on a philosophy that is tried and proven and has been successful in other places.

It has been said that the difference between an optimist and a pessimist is that the pessimist has more information. All it takes to overcome the genuine skeptic and put him or her on the side of right is some credible evidence and the opportunity to participate. The best place for a skeptic is on the quality improvement team, preferably in charge of something. Early exposure to some management training in quality, such as management college, is a good idea.

There are other people pretending to be skeptics who just don't want to do anything or don't want to step out into something new. These are not skeptics; they are just bullheaded and lazy individuals. No amount of evidence or thought is going to convince them. They will turn around when it is made clear to them that they are the only ones in that parking lot and when the boss obviously wants it the other way.

Now in most cases those who appear to be against quality improvement really are not against it. They have been misunderstood. I am always getting aimed at some senior person in a company who the coordinator or one of our counselors thinks is a roadblock. In practically every case, it is an erroneous assumption. They appear to be hanging back because they are trying to figure the best way of implementing quality improvement in their areas of responsiblity. Actually, they will do very well.

Always assume that people are vitally interested in the quality improvement process. They will act to fulfill your confidence. Don't put people through a continuous litmus test on their conviction. No one knows for sure what is going on in other people's heads. Assume the best and that is usually what happens.

If you do run into a real case of brain damage on the subject, just call your friendly quality consultant for help.

Question 94

You've suggested that for all the years you've been teaching the formulas to improve quality in this country, management hasn't listened and hasn't yet gotten the point. What's the single most important reason why management hasn't learned yet?

Answer

If I had to pick one single point, I'd say the hardest one for everyone has been that quality is the result of an operating policy, not a matter of the application of techniques. It has not been a major objective of management, during my lifetime anyway, to do things right. They have been raised to compromise everything in order to make the overall goal of producing profit. Only recently have they begun to realize that profitability is a function of customer satisfaction.

But don't let me mislead you. A lot of senior executives have gotten this point and are doing something about it. The percentage gets smaller as we go down the organization until we reach the lower third where they have known this all along. The quality professionals, for the most part, have not recognized how important this comprehension is to them. They, too, still think it is a matter of applying techniques. But there is a discernible change in that area too. A younger, more aware group is beginning to take control of the profession.

Question 95 _____

As a CEO, you've made a "believer" out of me. What do I do once I have completed my quality improvement education?

Answer

Pretend you are the scout leader and the company is the troop. Determination, education, implementation: That is what the leader is most concerned about. It is necessary to show by example that you, personally, are determined that the company is going to change, that everyone is going to make "scouting" an integral part of their life. Put quality first on the agenda of the regular management meetings. Serve as the chairman of the steering council. Get actively involved. Read the Fanatic's card and put those principles to work. Be careful to keep in close contact with those who are managing the process for you. Ask your quality improvement counselor how things are going, and do that regularly. It is easy for a process to get sidetracked if people get off onto ego trips or making little adjustments in the material or the flow of things.

Make certain that the educational programs are being utilized properly, that all individuals are having the opportunity to understand their personal roles in causing quality improvement. Visit some of the internal classes and encourage other senior executives to do the same. Ask that "show and tell" sessions be held once in a while so you can hear some of the success stories.

Quality implementation is a career-long task that comes from insisting that the quality policy be met. Be prepared to encourage your quality professionals and any others who will be dealing with suppliers, with statistics, and with awareness programs. Quality has to become part of the woodwork, just like profit.

There are CEOs all over the place who can give you information on how to do a good job in the role. Introductions to a couple of them would be easy to arrange. They lead but do not dominate the process. People have to embark on the quality mission because they want to do it, not because the boss forces them.

Question 96 _____

If you had to pick a famous management person for managers to model themselves after, who would it be?

Answer

Jesus of Nazareth. Seriously. He had very basic principles and stuck with them all the way. This is a man who never wrote a word, had a ministry that lasted only a few years, picked assistants who were the lowest of the low (in terms of their social class standing), and yet had a greater effect on the world than any other. He really was a man, not a magician or someone with secret powers. He performed miracles through the same faith that is available to each of us. That should give us a clue about conviction, commitment, and conversion.

The key to Jesus' teaching was the use of basic principles and the common everyday analogies he used to transmit them. For instance, He said, "Love one another." If someone came and said, "This man is telling lies about me," the response was, "Love one another." To the question, "What can we do about the communists?" Jesus would say, "Love one another."

Modern executives I admire are Harvey Firestone, Andrew Carnegie, Franklin Delano Roosevelt, Dwight Eisenhower, George Marshall, Jim Burke (the Chairman of J & J), Harold Geneen, Tom Willey, Roger Milliken, and several others. I like people who create without hurting others, who are positive, never alibi, and above all work to a set of principles that mean everything to them. The integrity of the individual determines his or her success in advance.

I read a lot of biographies and probably have 400 in my library. Also Executive Newstrack has autobiographical tapes they will send you. I think it is necessary to have a lot of role models and make a composite. By reading about successful people or listening to their autobiographies on tape, it is easy to determine that they come very much from the same mold. They are considerate of others and for the most part have a genuine faith. They see no need to be devious or unfair. They are people-oriented.

I meet many senior executives who think they are people-oriented but have annual reports with no pictures of people in them and continually make decisions based on everything but people. It is hard to snuggle up to numbers.

Epilogue: Quality Past, Present, and Future

The way quality has been dealt with over the past years was a good news, bad news story. The good news was that defects were contained to the extent that things could be made to work. The bad news was that the basic concept of quality declared that nothing could ever come out right and the best that could happen was containment. Companies never had any real intention of getting things done right the first time because it never dawned on them that it was possible.

Quality control concepts, techniques, and practices were developed around assumptions of the inevitablity of error and made no room for a defect-free situation. When one occurred, the assumption was that something had been missed and that it was a success due only to the inadequacy of the appraisal effort. Quality control was only applied to the manufacturing activity.

Management was given no opportunity to look at quality as something where the governing principle was anything other than pure CHANCE. Completely immersed in this thought, management abandoned its role in handling quality. It knew what to do about money and it could handle time, but quality was in the hands of the gods.

When I went to work on my first day at Crosley, I was overwhelmed with the knowledge that everyone else possessed. When I joined the quality society and began to read about the subject, I realized that I had stepped, by accident, into something that was very complex and was understood by only a few. It all sounded logical and seemed to make sense, but it was beyond me.

The essentials of quality management in the days of CHANCE were:

Quality is a desirable characteristic.

Quality is achieved by appraisal.

The quality performance standard is "acceptable quality levels." Quality measurement is by indexes and ratios.

As the years went by I began to realize that companies had no intention of meeting the requirements they promised their customers, and the customers didn't really expect them to be met. I was disillusioned and concerned. There were many bright, well-educated people in the profession and in management. Yet they accepted this situation as the way it had to be. At the same time they were overcoming all kinds of other challenges. Business went on, grew, and was profitable.

The most valuable members of the quality function were those who could work with the customers to develop the compromises that produced waivers and deviations. Suppliers did the same with us.

Management was measured primarily on things like "pounds out the door." Assembly lines were never to be stopped; things could be fixed later. White-collar work had nothing to do with quality, so quality became a step-child, ignored by all.

I churned along doing the things that were considered normal and even adding to these developments. When I went to Martin in 1957, I was helping to bring the modern quality assurance operation to the surface. We had great procedures, the complete support of the government quality people, and a grateful management. However, like everyone else, we had a lot of nonconformances and deviations. We didn't plan to get everything right. We all knew that was impractical.

Three events occurred to me that made me realize I was going to have to face reality and pay attention to the nagging feeling inside me. I just couldn't get away from the thought that it didn't have to be like this.

First, a casual conversation with a veteran of the business made it clear to me that all of the new creative quality assurance and reliablity things we were so eagerly implementing had not changed things one bit and weren't likely to do so. The percentage of nonconforming material, the number of material review actions, the customer complaints, the whole ball of wax had not been any different during his thirty years in the business. We had not improved. We had just gotten more expensive.

Second, in putting together a list of problems for one weapons system program, I realized that the list was almost identical to one I had created at a different company. I was beginning to feel like Sisyphus, who was condemned to roll a rock almost to the top of the hill, only to see it career back to the bottom.

Third, senior management, in a fit of pique, challenged me to deliver products that didn't have anything wrong with them, to do it right the first time. That seemed to rip the scales from my eyes and make me realize that the reason no one improved much was because they didn't think it was possible. A person's greatest insights always seem obvious upon reflection. I can remember changing my swimming time to afternoon when I realized that the water was warmer then. I had been going in the morning because that was when I had always gone.

Out of all this, of course, came the concept of zero defects, which says that management receives what it asks for, and that if it wants a certain percentage of nonconforming material, then it will receive that percentage. However, if management encourages prevention and helps people, they can receive defect-free products and services.

This revelation, which came in 1961, was not received well by the thought leaders in the quality field. I have documented all of this in several books and articles. But it was a shock to me. I always thought improvement was something that everyone wanted. That is not the case.

It wasn't until the Japanese invasion of the 1970s that acceptance of this thought began to be more general. It was driven home because that is what the Japanese were doing.

One could actually purchase a TV set that worked and did not require service every whipstitch. The industry screamed about "price dumping" but that was not the case. While the Japanese were taking quality seriously, the Americans were doing business as usual. The automotive and steel people learned nothing from the demise of the consumer electonics industry.

Today, quality is an important part of executive thinking. Popular books such as *In Search of Excellence* have raised the level of consciousness and let people know that the situation didn't always have to remain the way it had been. All of this brought on a burst of energy, harder and better efforts to control quality, with predictable results. I could see as I began to get feedback from *Quality Is Free* and as I set up my own quality improvement business that senior management was becoming aware that they could have a CHOICE as far as quality was concerned. They could do well if they wanted to, but it required policy changes and a different way of operating.

They accepted the new essentials of quality management, which I call "The Absolutes:"

Quality is defined as conformance to requirements, not as goodness.

Quality is achieved by prevention, not by appraisal.

The quality performance standard is zero defects, not acceptable quality levels.

Quality is measured by the price of nonconformance, not by indexes.

Changing a company to operate according to these concepts requires more than posting them on the walls. Executives were aware that everyone had to come to an understanding and that they had to make it happen. As a result, my associates and I developed a multilevel system of education for training individuals as well as companies in quality improvement and the implementation process. Executives and managers could be taught in classrooms and the other 97 percent of a company's work force could be taught at the workplace using the materials we developed. Instructors were taught how to deliver it.

What is never realized is how long it takes to absorb such a change. I read about companies doing wonderful things with quality and recognizing that it takes several years to completely understand why things work and why they don't. Unfortunately the things that do the least good often get the most attention. Few golfers, for instance, spend much time practicing putting. Yet that is the most important stroke in the game. A regulation par round would be 50 percent putts and 50 percent all the other shots.

The tools of quality control are very useful, and they are readily available. But they don't bring about much change. They have to be put in perspective. Understanding and meeting the customer's requirements is the important factor.

A lot of companies are doing something about quality. Probably 20 percent are on the right track; 40 percent are doing something but not doing it properly; and 40 percent are waiting for something to happen, like a law, that will protect them.

Europe has come alive in the past two years. China, Taiwan, the Philippines, and several other far eastern nations have been slower to get involved with managing quality. They are still putting their efforts into line control. That is why their main products are mass production items.

As executives began to realize that they could really have a CHOICE, they started to take charge of the quality part of their businesses. In the next few years there will be many changes, and it can't come too soon.

We live in a world economy, and in the future quality will not just be something that is nice to have. It will be the necessary price of admission to the market. Customers will no longer be impressed to receive what they ordered. They will consider it their right. So companies who have perfected the art of explaining why things are not as promised will have no future.

Management will have to CAUSE quality to be part of the company culture, a normal event. Going from CHANCE to CHOICE to CAUSE in a few years shows what can happen when reality strikes.

What holds efforts back is the rather naive notion that quality is something easy to do once the mind is set on it. Authors investigate what companies have and have not done, write about it, expound on it, and never really understand what is involved in changing. They would never dream of doing this about something they viewed as complicated. They think, for instance, that any quality improvement process will succeed if one sticks to it. They find themselves recommending ways of doing things that have failed for years. It reminds me of the French and British general staffs in World War I who refused to accept the existence of the machine gun.

The amount of dedication and effort necessary to bring a company to world-class quality is parallel to what it takes to become a world-class athlete. Complete concentration on the subject by the most senior concentrators is a routine part of that. Crime is not prevented by the police department, although having one makes criminals think harder. Crime is prevented by eliminating the need for it and the acceptance of it. The same is true with quality. Laws and discipline are helpful, but integrity and fair dealing are the keys.

Companies will not learn these things by themselves. And the lower levels of an organization will not change things on their own. I have seen dozens of companies struggle and fail because the top executives would not insist on change. They let others muddle along with it. Eventually they become desperate enough to do something about it and begin a culture change that is not dependent on technical solutions.

The customer is the key. It is when the customer is happy that a company is truly doing well, not when the management is happy.

Guidelines
for
Browsers

I think everyone learns more in the past five years than in all the previous years in any field.   3

Attitudes change when a business's culture or working environment is changed, not until.   3

When it is a pleasure to come to work because the requirements for quality are taken seriously and management is helpful, then attitudes change permanently.   3

Teaching people, leading people, showing people, providing tools—everything loses meaning if employers, customers, and suppliers feel that management is not walking like they talk.   4

People should spend their time improving the quality process, rather than juggling it around to meet their feelings of the day.   5

Don't set up false detours or special arrangements. Learn what right is and do it that way all the time. Then people will have something they can trust.   6

When management encourages procedural Band-Aids, employees lose confidence in them and in the process.   6

Change should be a friend. It should happen by plan, not by accident.   7

Good ideas, based on solid concepts, have a great deal of difficulty in being understood by those who make a living doing things the other way.   7

In a true zero-defects approach there are no unimportant items.   9

The zero defects idea never died. It just had a long gestation period.   10

No idea was ever accepted right out of the chute.   10

The first phase of change is developing conviction....The second phase is commitment....The third phase is conversion.   *15*

Laying out a specification that says something undesirable is acceptable does not make it right. The requirement has to be something that delivers to customers what those customers think they are buying.   *17*

The purpose of quality is not to accommodate the wrong things. It is to eliminate them, to prevent such situations.   *17*

When you have an environment where nothing is certain, where almost anyone can make changes or write deviations, then the output is uncertain at best.   *19*

It is hard to get people interested in improvement of any kind if they perceive it as a threat to their authority or lifestyle.   *19*

When it comes to quality, the company that teaches its management about prevention and makes that a vital part of the day-to-day upper-level conversation of the company will forge ahead eventually.   *27*

Listen to everyone; seek out your customers and interrogate them. Don't fall into ruts. Do what you do thoughtfully, but don't go to sleep. Someone may be out in the garden digging up your treasure.   *27*

You never put development people in charge of production. They get too used to changing things whenever they feel it is necessary.   *28*

Quality is not a matter of having some superknowledge in some supergroup. It is a matter of managerial integrity. Either the requirements are taken seriously or they are not taken seriously.   *29*

There is too much emphasis on the ineffective aspects of quality improvement. Many people still think it is a technical problem, not a people problem.   *29*

American management does not work hard on prevention. They will strive until they are exhausted to overcome some problem, but prevention is thought to be for sissies.   *30*

[American] policy has been to make 1200 in order to have 900 that we could sell. [The Japanese] learned to make 1200 to have 1200 to sell.   *30*

The quickest way to turn around is for the management of companies to take charge of the education of their people and help them to learn the necessary things.   *31*

As it is, business schools are actually creating business for consulting firms by teaching quality wrong.   *31*

[Schools] can support and reinforce the ideas, but the values that underlie our message and methods should come from the home.   *31*

When the corporations in this country tell the schools that they are not satisfied with the way quality and other things are being taught, things will change.   *32*

Changing the national process is going to take a while, but demanding that it change is always the way to start.   *32*

U.S. managers, with a few notable exceptions, don't give the job their full, undivided attention day and night. They do not look for ways to finalize problems and opportunities forever.   *33*

What made the Japanese do well at quality was the dedication of senior management. They realized that they could not mount a worldwide service department to keep their products operating. They had to build them right the first time.   *33*

I am frequently asked why I do not recognize that there are certain "natural laws" that keep everything from being right. *34*

Everything must be questioned continually. I learn every day. I would rather have U.S. managers than any other, but they need to learn how to get things done, and then do some more. *35*

If that nation of a billion business people [China] ever gets turned loose to use their natural talents, it will be a tougher world in which to earn a living. *36*

Momentum is not what keeps the quality process moving or makes it a permanent part of a company's structure. Necessity and success push it along. Once people learn to work this way, they will not want to give it up. *38*

If something is worth doing, it will continue regardless of temporary situations or setbacks. And the best way to keep the bottom line attractive is to learn to prevent loss and waste. *38*

Good managers, whether personal or professional, do not change policy or objectives because of something that jumps up in front of them. Strategy changes are continual; policy ones are not. *39*

At this time most of the country is still in darkness, fighting the quality battle the old-fashioned way, with excuses. *40*

People get caught up in misunderstandings and in their own need to drive the wagons over the cliff personally. *41*

The thought that will do you the most good is "prevention for the purpose of causing defect-free work." That will reverse the way a company normally operates. *45*

Neither your company, nor any other company for that matter, has ever put out an advertisement that says the products or services produced will contain errors and defects. *45*

There is no reason for not taking requirements seriously, and people will do that when it is what the organization believes. *46*

So if you can take away an understanding that it is very much to the benefit of your stockholders, employees, customers, and suppliers to begin a tradition of doing things right the first time, this will have been a successful experience for you *and* for them. *46*

Improving quality requires a culture change, not just a new diet. *47*

If we make our profit goals, but don't pay our bills, then we have not met our profit goals. If we deliver on time, but the product has defects, we have not delivered on time. If we meet our safety objectives, but damage somebody, we have not met our safety objectives. *49*

I used to get a lot of free lunches by betting general managers that I could find someone violating a safety rule during our tour of their facility. It was always the general manager. *49*

The thought of error being inevitable is a self-fulfilling prophecy. If you think it has to be that way, it will be that way. *51*

A person is either an entrepreneur or not. That person is born with the intent and learns how to implement—if he or she can recognize this entrepreneurial talent. *56*

An entrepreneur is anyone who can develop a plan, explain it to others, get them to eagerly help, and at the same time keep a consistent pressure on the enterprise... *59*

The difference between an entrepreneur and an executive is that one is customer-oriented and one is management-oriented. The executive runs what the entrepreneur brings together.   *59*

Requirements are answers to questions and the agreements that result from those answers.   *61*

So we need to keep "requirements" in proper perspective. It is we who are the masters, not the requirements. They serve to mark an agreement between people and should take whatever form is necessary. They must be respected and never altered except by agreement between those who created them.   *62*

We have to be able to count on each other doing what we have agreed to do.   *63*

Everything that happens in any operation today happens because of requirements that exist. They may be written down, they may be verbal, they may be traditional, they may have just been spoken a moment ago.   *63*

Zero defects is a symbolic way of saying "do it right the first time."   *63*

The real risk takers, like those who climb mountains or go into space, make sure that things are right the first time, and they do it before going.   *64*

The key to "risk taking" is taking time to lay out the requirements, to get clear on everything that is known. Some things are not known and have to be planned for, but if the rest of an operation is well based, the unknowns become something that can be dealt with.   *64*

When people are too lazy or preoccupied to work out the requirements, then they are taking risks.   *64*

If an organization is really to be run properly, eveyone has to understand the purpose of it. Everyone has to know the charter of the organization, and they have to understand their personal role in making it all happen. 65

Zero defects is doing *what* we agreed to do *when* we agreed to do it. It means clear requirements, training, a positive attitude, and a plan. 65

The conventional notion of risk taking is that it involves leaping off into the unknown. People who leap off into the unknown disappear into that same unknown. 65

Zero defects is the result of thinking things out. 66

A great many discrepancies occur merely because they are expected to happen. That expectancy becomes so routine that people spend their time learning how to fix rather than prevent. 66

One of the most cynical statements imaginable is known as Murphy's Law. 66

I would not want to work in an organization whose objective was less than zero defects. 68

We have to want to achieve something that improves the satisfaction we have with life. 69

There is a great deal of frustration and an enormous lack of satisfaction for the great masses of working people. Much of this is due to their determination that the specific tasks they do are not important or meaningful. 69

A company's purpose is to give people worthwhile lives by providing the opportunity for meaningful work, a decent living, and an opportunity to make a contribution to others. 69

I figure that every other person in service companies spends 100 percent of their time doing things over, chasing after data, or apologizing to someone.   72

All work is a process.... Whenever anything goes wrong in one area, the shock waves are felt throughout the organization. That is why it is so important for everyone to be involved in efforts to do things right.   72

If quality is to be managed, it must have a meaning that is manageable.   74

"Quality" as used in quality control and quality assurance has always meant goodness because people were permitted to make value judgments every day. "Goodness" was really what it was all about. However, when the thought of conformance and nonconformance begins to permeate the operation, "goodness" suddenly becomes inadequate.   74

All requirements come from the customer in one form or another, because with no customers there is no business.   75

All words have specific meanings. When we use them to communicate, we have to take time to make certain we are all singing the same song.   76

The proper usage of words in communication is very important in an organization. When imprecise words are used there is confusion and dilution of effort. The proper requirements make everything possible.   76

More "systems" are not needed. Harder work and a more serious application are what require attention.   78

The quality mission involves helping companies develop an overall strategy that will serve them for years, and this must be continually updated.   80

Management installs programs to solve problems on an ad hoc basis or to move the company along instead of adjusting the company culture to meet the realities of the marketplace. *81*

Management has to understand what the system is and has to take part in making it a living, breathing part of the operation. *81*

The success of the quality improvement process does not depend on any "evangelical" powers possessed by the quality experts. It depends on education and implementation conducted in a serious and methodical way. *82*

You have to be careful because management just loves to find packages that replace thought and original work. *83*

It always bugged me that someone I didn't choose got to make decisions about my career. *86*

Building up a large data bank in one's head provides a source of understanding about situations. Then the thought process will work its way around to something practical and usable. *91*

The quality improvement process is progressive. One doesn't just go from awful to wonderful in one leap. *94*

It is dishonest to advertise defect-free stuff and then not plan to deliver it. *97*

Witnessing is not just standing up in front of everyone and being evangelical. The populace wants to see if management "walks like it talks." *102*

If management caves in to the pressures of time and money, then quality immediately crashes to its place as number three. *102*

Each and every person in the organization must understand his or her personal role in making quality happen. *102*

The buying operation has the responsiblity of making certain that the requirements are clear, the supplying organization has the responsiblity of making certain that they can deliver exactly that, when needed. *103*

Employees need to understand where their work goes. They need measurements in order to keep track of how they are doing. *103*

One of the perils that comes from an organized attack on some problem is that people will begin to feel that the system will take care of it all. *103*

Satisfy the customer, first, last, and always. *104*

I think that trying to torpedo the competition is not part of the business world. There is plenty of room for everyone. *105*

Management commitment is the willingness to give away something you cherish, something very personal, in order to improve the quality of other people's lives. *106*

Most real changes start somewhere in a company and spread around because they are worthy. They do not come out of the boardroom. *108*

If we waited for everyone to be at the same level, we would never get anywhere. *109*

A quality improvement team is not a SWAT squad that has to go solve problems. Quality improvement training is education of the team members, coordination between operations, and leadership of the process. It is a living entity that survives its members. It is not a ritual, it is leadership. *109*

It only takes one person, one division, one group to change the whole company. *112*

If everyone back at headquarters is running around wearing white robes and burning incense because of their conversion, but the same old stuff is still coming off the production lines, it is understandable that the sales force will not take it all seriously. *112*

Relationships are what business is all about, and there are only two that matter: the one with customers and the one with employees. *112*

If you think about it, most of what happens in the quality improvement process, as well as in the rest of running a business, is done by means other than face-to-face contact. *115*

The process of work in all companies is service. *116*

We have to learn to measure the process of work as it moves along, not wait until a tangible product appears and then swarm all over it. *117*

Everything can be changed for better or worse, and it will change by itself when the time comes. So it benefits management to cause proper changes rather than just waiting for events to run their course. *119*

Things can be whatever the leader wants, but it has to be spelled out. Every day is a brand new day. The problems that have been around for a long time don't have to be there. *121*

The key to everything in this employer-employee relationship is providing a clear job description and measurement system that both parties understand. *122*

The keystones of science are integrity and measurement. That is what the quality improvement process is all about too. *122*

Being creative does not mean grabbing elements out of the blue and slapping them together in the blind hope that the result will be a viable new product or a new twist on an old product.   *122*

Few people in a company, if any, understand quality or can talk about it.   *124*

An extensive up-front audit is a way people postpone doing something truly useful.   *124*

I start with two beliefs. First, those asking for help probably don't know much about it or they would be fixing it themselves. Second, there is someone there who does know and no one will listen to that person.   *125*

Someone has to be responsible for everything—everything—or it won't get done.   *126*

It is the worst kind of egotism to take something that has been 25 years in the creation and to modify it on the basis of an assumption of understanding and in an attempt to use it for personal ends.   *127*

It is a mistake to get involved in anything that sends out the message that you are not serious about zero defects.   *129*

We do not make change without serious thought, but there is a time and a way to do it.   *131*

Quality has always been the most negotiable of the "schedule, cost, quality" triumvirate.   *132*

There are no failures in introducing quality improvement. No one ever gets worse and no one ever doesn't improve.   *133*

Creativity flows freely when understanding is universal.   *134*

Success is a matter of return on effort.   *134*

When checking progress, the good news is finding out what needs something done about it. The bad news is that something has to be done about it.   *135*

The true growth factor in quality improvement is the reduction of errors, such as in process improvement, and in getting more work out of people because they are only doing things once instead of several times. *135*

There is a Quality Revolution going on and the quality people are missing it because they have their eyes on what goes out the back door, whether it's a product or a service. *140*

Cost reduction is a byproduct of quality improvement, but people will not work hard to reduce costs unless their lives get easier along the way. *140*

There is a marvelous opportunity and a bright future for those who learn how to manage quality in a manner that makes it an obvious asset to the company rather than a "necessary evil." *142*

People are thinking, caring beings, and they can tell whether you respect them or not. They work for appreciation and the sense of accomplishment they get from doing the job well. They do not work for money. They need it, and it is important, but money is a lousy motivator. *143*

It is up to the management to arrange that the proper recognition be given to those who accomplish and the proper reeducation or reassignment be done with those who are having problems. *143*

Proper understanding and proper communication reduce hassle. Most feelings of frustration in any relationship come from not being heard. *147*

Not having a common understanding of quality puts more pain into an organization than anything else I have ever known. *147*

The result of agreeing to this small exception here and that medium-sized compromise there and this larger but understandable problem somewhere else is a general deterioration of quality.   *148*

Perform in accordance with the requirement, or change it offically to what the company and the customer really need.   *148*

Unfortunately, most quality professionals think they are responsible for quality and so become unwitting accomplices to their own frustrations.   *149*

The main reason for doing quality improvement is not cost elimination, it is to satisfy customers and get the company in better shape.   *149*

Employees will see more hope than fear in a well-planned, well-managed, quality improvement process.   *150*

When everyone understands what they are looking at, there is no negative.   *151*

If we don't measure, then we cannot communicate.   *151*

It's best to look at measurement as a normal part of operations, something to be taken as part of the scene.   *151*

Any good thing can be turned negative.   *152*

We have to be honest with people if we want them to be honest with us.   *152*

Creative people become discouraged when the things they create are not implemented properly.   *153*

We need to concentrate on establishing clear requirements and then meeting them routinely, not on explaining how the wrong mixture is really okay.   *155*

The customer and the supplier have to have a relationship based on mutual need.... Even bank robbers need somebody to stick up.   *156*

There are no marketing advantages that last very long. Something always comes along to serve as a replacement or substitute, or to eliminate the need entirely.   *157*

Getting requirements clear between supplier and purchaser is one of the most important parts of the quality improvement process.   *157*

You can't teach executives inside their own company. They just will not take any of their peers seriously.   *161*

Having the big boss on your side only provides protection as long as what you do is useful.   *162*

There are no politics in quality unless someone is threatened with failure and public exposure.   *163*

It is very difficult to deal with a subject when people cannot agree on the meaning of key words.   *164*

If you get your board and the other senior executives to use a common language of quality and to recognize their personal role in making it happen, everything will come out just the way you want it.   *165*

Management has three choices: (1) to learn to live with the delay and inconsistency that comes from the boat getting hung up on a regular basis; (2) to spend a lot of money and effort to put more water in the stream, which increases overhead and reduces profitability; (3) to clean out the river bottom by removing the obstacles to defect-free operation.   *166*

Don't adapt the quality improvement process to the culture, change the culture to conform to what is best. Learn from the past, but don't live in it.   *168*

The difference between discussing and arguing is whether the participants are using facts or opinions.   *168*

Be pleasant and cooperative, but don't let anyone beat up on you. Life's too short for that.   *169*

In most "people relationships," an honest, open chat about specific items will resolve most conflicts.   *169*

The true skeptics are usually those who only want more than an emotional "feel" in order to be convinced.   *170*

All it takes to overcome the genuine skeptic and put him or her on the side of right is some credible evidence and the opportunity to participate.   *170*

Always assume that people are vitally interested in the quality improvement process. They will act to fulfill your confidence.   *171*

[Managers] have been raised to compromise everything in order to make the overall goal of producing profit. Only recently have they begun to realize that profitability is a function of customer satisfaction.   *171*

Quality implementation is a career-long task that comes from insisting that the quality policy be met....Quality has to become part of the woodwork, just like profit.   *172*

People have to embark on the quality mission because they want to do it, not because the boss forces them.   *172*

The integrity of the individual determines his or her success in advance.   *173*

I think it is necessary to have a lot of role models and make a composite.   *174*

It is hard to snuggle up to numbers.   *174*

A person's greatest insights always seem obvious upon reflection.   *179*

We live in a world economy, and in the future quality will not just be something that is nice to have. It will be the necessary price of admission to the market.... So companies who have perfected the art of explaining why things are not as promised will have no future. *181*

Laws and discipline are helpful but integrity and fair dealing are the keys. *182*

Index

Absolutes, 50–52, 180
Accidents:
  safety and, 49
  on space shuttle, 27–29
Acupuncture, 38
ADEPT block, 118
Airline industry, 17–18
American Society for Quality Control, 164
Appreciation, 142–144
Attitudes toward quality, 3–4
Audits:
  for quality improvement process, 123–124
  of successes, 133–135
Automobile industry, 23–27, 167
  General Motors, 21–23
Awards, 143, 162
  Beacon, 143, 144

Beacon Awards, 143, 144
Boards of Directors, 163–165
Borel, Georges, 35
Bosses, disagreements with, 168–169
Boy Scouts, 76–77
Britain, production in, 37
Business schools, 31, 83
Businesses:
  audits in, 123–124
  company politics in, 161–163
  profits from quality improvement process in, 135–136

Businesses (*Cont.*):
  service, quality improvement process in, 116–117
  skeptics in, 170–171
  small, implementation in, 117–121
  (*See also* Corporations)

Case histories, 88–90, 93–94
Challenger space shuttle, 27–29
Change:
  in cultures, 165
  organizational, 15–16
  planned obsolescence and, 7
Chief executive officers (CEOs), 4, 172–173
China, 36–38, 150
Chrysler Corporation, 24, 26–27
Churchill, Winston, 109
Commitment, 15–16
Communications, 115–116
  of requirements, 160–161
Companies (*see* Businesses; Corporations)
Company politics, 161–163
Computers, 108
Conformity, 60
Congress, NASA oversight committee of, 27–29
Consultants, 129, 146
Conversion, 16
Copperfield, David (magician), 83–84
Corporate culture, 165–168

Corporations:
  audits in, 123–124
  Boards of Directors of, 163–165
  chief executive officers of, 172–173
  company politics in, 161–163
  price of lack of quality in, 71–72
  quality improvement process in
    departments of, 107–108
  skeptics in, 170–171
  vertical integration within, 159
  "wellness" of, 4
  (*See also* Businesses)
Costs, 132–133
  of nonconformance, 149–150
  prices versus, 73
  of quality, 124
  quality improvement process tied to
    reductions in, 140
Crosby Complete Quality Management
  System, 40, 79, 130
Customer requirements, 76
Customers:
  quality requirements of, 154–155
  requirements originating in, 75
  satisfying, in quality programs, 104
  services to, 155–156
  as suppliers, 158–159
  of unique products, 156–158

Davies, Brian, 92
Defects (*see* Zero defects)
Defense, U.S. Department of, 45–46,
  105
DeLorean, John, 24
Deming, Dr. W. E., 78, 79
Directors, 163–165
Discharging employees, 144–145
Discipline, 142
Drucker, Peter, 82–83
Dunleavy, Tom, 13

Economics of quality, 52
Economy, recession of 1982 in, 39

Education:
  of chief executive officers, 172–173
  company politics and, 161–163
  of employees, in quality, 102
  in Japan, 31–32
  in quality improvement, 109–110,
    180–181
  (*See also* Training)
Electric utilities, 34
Employees:
  discharging, 144–145
  education of, in quality, 102
  field staff, 112–114
  job security of, nonconformance tied
    to, 149–150
  motivation of, 104–105
  performance appraisals of, 85–87
  quality assurance professionals,
    139–140
  quality circles among, 14
  in quality improvement teams,
    108–110
  sales force, 114–116
  service, 116–117
  in stores, training of, 121–122
  "tag men," 167
  training of, 130–131
  turnover of, 147–149
Entrepreneurialism, 56–60
Error cause removal, 70, 107
Ethics, 31
Excellence, 74, 76
Executive Newstrack, 173–174
Executives, 173–174
  education of, 161

"Family council," 113
"Fanatics" cards, 101–104
Field staff, 112–114, 156
Fitness for use, 75
Floyd, Ray, 160
Food industry, 16–17
Ford, Henry, 85
Ford Motor Company, 24, 26

Four Absolutes, 50–52, 180

Geneen, Harold, 58, 148–149, 162
General Electric, 116, 124
General Motors, 19, 20, 26–27, 116
  Corsica and Beretta cars from, 21–23
  implementation of quality improvement process in, 111
  rebuilding of, 25–26
Girl Scouts, 76–77
Goals, 107
Golf, 47, 67–68, 160–161, 181
Goodness, 74
Graphs, 150–152
Grau, Diter, 29
Grids, 94–95

Halbersham, David, 24
Halpin, Jim, 86–87
"Hassle factor," 86
Health care industry, 72
Hong Kong, 37

IBM Corporation, 116
Implementation of quality improvement process, 3, 70–71
  consultants in, 129
  employee turnover and, 147–149
  fourteen steps for, 106–107
  in parts of organizations, 111–112
  in small businesses, 117–121
  successes in, 133–135
  suppliers in, 111
  training in, 108–111
  of zero defects, 69–70
  by management, 79
Innovation, 153–154
Integrity, 74
Internal suppliers, 159
Inventory control, just–in–time system of, 84–85
Ismay, Lord, 109

ITT, 13, 58, 115, 148–149
  executive education at, 32
  executive recognition in, 162–163
  implementation of quality improvement process in, 70–71
  quality councils in, 35

Japan, 179–180
  auto industry in, 25
  education in, 31–32
  just–in–time inventory control in, 85
  management in, compared with U.S., 32–36
  national character of, 35
  quality in, 29–30
  zero–defects concept in, 9
Japan Management Association, 33
Jesus Christ, 173
Job methods, innovation in, 153–154
Job security, 149–150
Juran, Dr. J. M., 79
Just–in–time inventory control, 84–85

Kobyashi, Dr., 9

Labor costs, 36
Lincoln, Abraham, 110
Lister, Joseph, 8
Lowrie, Raymond, 6

McClellan, George B., 110
McDonald's Restaurants, 122
Management, 171
  audits of, 123–124
  commitment to quality by, 102
  in corporate culture, 166
  models of, 173–174
  programs understood by, 81
  quality improvement process opposed by, 145–146
  responsibilities of, 105

Management (*Cont.*):
  role in quality of, 13–14, 163–164
  in U.S. and Japan, 32–36
  working around, 108
  zero defects to be implemented by,
    79
Management by objective (MBO),
  82–84
Manufacturing, 19
  innovation in, 153–154
  price of lack of quality in, 71–72
  quality assurance departments in,
    140–142
Martin (firm), 119–121
  field staff in, 113–114
Maturity Grid, 94
Measurement:
  graphs for, 150–152
  in research and development,
    122
Milliken (firm), 110–111
Motivation, 104–105, 140
Murphy's Law, 66–67

National Aeronautics and Space
  Administration (NASA), 27–29,
  149
National character, 35
Nelson, Bill, 27
Nicklaus, Jack, 160–161
Nippon Electronic Corporation
  (NEC), 9
Nissan (firm), 24, 25
Nonconformances, 74, 148
  correcting, 166
  job security tied to, 149–150
  price and cost of, 73
Norman, Greg, 160
Nuclear energy industry, 62–63

Objectives in management by
  objective, 82–83
Obsolescence, 6–7

Organizations:
  change in, 15–16
  implementation of quality improve-
    ment process in parts of,
    111–112
  quality assurance departments in,
    140–142
  quality improvement process in
    departments of, 107–108
  suppliers, 159–161

Pacific rim nations, 36–38
Palmer, Arnold, 160
Pareto principle, 9
PCA (Philip Crosby Associates, Inc.),
  131
Perfectionism, 75–77
Performance:
  employee appraisals for, 85–87
  minimum requirements for, 152
Personnel (*see* Employees)
Petroleum crisis, 24–25
Planned obsolescence, 6–7
Pontiac, 24
Prices, costs versus, 73
Processes, programs versus, 139
Productivity, 74
Profits, 135–136
Programs, processes versus, 139

Q–21 (Quality in the Twenty-First
  Century) videotapes, 128–129,
  132
Quality, 177–180
  attitudes toward, 3–4
  conformance with requirements, 58
  costs of, 124
  culture of, 165–168
  differing concepts of, 164
  economics of, 52
  Four Absolutes of, 50–52
  implementation of, fourteen steps
    for, 106–107

Quality (*Cont.*):
  measurement of, graphs for,
    150–152
  as negotiable, 132–133
  price versus costs of, 73
  safety and, 49
  in stores, 121–122
  synonyms for, 74–75
Quality assurance departments,
    140–142
Quality assurance professionals,
    139–140
"Quality Bowl" game, 134
Quality circles, 13–14
Quality consultants, 129
Quality control, 50, 177
  Challenger space shuttle accident
    and, 27–28
Quality councils, 107
Quality improvement process:
  adaptation within, 126–127
  in airline industry, 17–18
  audits for, 123–124
  change handled by, 7
  for chief executive officers, 172–173
  company politics and, 161–163
  customers in, 155–156
  education, 109–110, 180–181
  employee turnover and, 147–149
  in food industry, 16–17
  implementation of, 70–71, 129–130
    in parts of organizations, 111–112
    (*See also* Implementation of
        quality improvement process)
  in individual departments, 107–108
  innovation and, 153–154
  just–in–time inventory control and,
    84–85
  losses and, 38–39
  management opposition to, 145–146
  management understanding of, 82
  management by objective in, 82–84
  nonparticipants in, 144–145
  in organizational structure, 140–142
  positive examples of, 20–23

profits from, 135–136
quality assurance professionals in,
    139–140
in research and development,
    122–123
reward and discipline in, 142–144
sales force in, training of, 114–116
in service firms, 116–117
successes in, 133–135
Quality improvement teams, 108–110
Quality Man, The (film), 91–93,
    163–164
Quality in the 21st Century project,
    20–21

Recognition, 142–144
  of executives, 162
Reliability, 50, 55–56, 153–154
  failure built into, 67
Reports, 141
Requirements, 61–62, 76
  changing, 148
  communication of, 160–161
  conformity with, 60
  from customers, 75
  customers' perceptions of, 154–155
  establishment of, 62–63
  minimums for, 152
  quality as conformance to, 55, 58
  for unique products, 157
Research and development, 122–123
Rewards, 142–144
Ring of Quality, 162
Risk taking, 63–65
Rude bosses, 169

Safety, 48–49
Sales force in quality improvement
    process, 114–116
  field, in implementation, 112–114
Sampling, 5
Santayana, George, 5

Schedules, 132
  Challenger space shuttle accident
    and, 28
  in management by objective, 83
Schweitzer, Albert, 107
Semiconductor industry, 34
Service firms:
  price of lack of quality in, 71–72
  quality improvement process in,
    116–117
Skeptics, 170–171
Small businesses, implementation of
    quality improvement process in,
    117–121
Smoking, 15, 119
Soldering, 52–53
Sony (firm), 33
Specifications:
  customers' perceptions of, 154–155
  in food industry, 16–17
Staff (*see* Employees)
Statistical process control, 14, 80
Stores, quality in, 121–122
Successes, 133–135
Suppliers, 119–120
  as customers, 158–159
  decision makers in, 159–161
  in implementation of quality
    improvement process, 111
  just-in-time inventory control and,
    85
  in quality programs, 103
  of unique products, 156–158
Systems integrity, 7

"Tag man," 167
Technology, quality improvement
    process in, 122–123
Tennant (firm), 3

Toyota (firm), 25
Training:
  of employees, 130–131
  of field staff, 113–114
  in quality, in stores, 121–122
  in quality improvement, 180–181
  of quality improvement teams,
    108–110
  refresher courses in, 110–111
  of sales force, 114–115
  (*See also* Education)
Troubleshooting, 124–126
Truman, Harry S., 75
Turnover of employees, 147–149

United States:
  management in, compared with
    Japan, 32–36
  in World War II, 118

Values, 96–97
Vertical integration, 159
Von Braun, Wernher, 52

Wellness, 4
Witnessing, 4, 102
Workers (*see* Employees)

Zero defects, 7–10, 179
  as goal and as reality, 66–68
  in golf, 67–68
  implementation of, 69–71
  risk taking and, 63–65
Zero–defects day, 70, 105, 107
Zimbalist, Ephraim, Jr., 89
Zoeller, Fuzzy, 160

About the Author

Phil Crosby has been a prime mover in the quality business for 36 years. For fourteen of those years he viewed the quality challenge from within the corporate world as a vice president at ITT. Now chairman of Philip Crosby Associates, Inc., he is one of our most highly respected and sought after international quality management consultants and educators. In addition, Crosby is among the best-selling authors in the field, with such outstanding quality and management works to his credit as *Quality Is Free*, *Quality Without Tears*, *The Art of Getting Your Own Sweet Way*, *Running Things*, and *The Eternally Successful Organization*, all published by McGraw-Hill.